Unlock the Secrets to Mastering Welding Techniques

Jamesz H. Ward

Funny helpful tips:

Rotate between different subjects; it prevents cognitive fatigue and keeps the reading experience fresh.

Stay updated with literary journals; they offer reviews, critiques, and discussions on current literary works.

Unlock the Secrets to Mastering Welding Techniques : Discover Proven Strategies and Techniques to Excel in the Art of Welding for Beginners

Life advices:

Engage with autonomous vehicles; their potential to transform transportation requires understanding their tech and implications.

Prioritize debt management; a healthy balance sheet attracts investors and partners.

Introduction

In this book, you will find essential information and guidance to help you get started in the world of welding and fabrication.

The guide begins by highlighting the equipment and necessary items for welding. It emphasizes the importance of a welding helmet and provides insights into selecting the right helmet lens for optimal visibility and protection. Tips for keeping your helmet clean are also provided to ensure clear vision during welding.

The guide covers the importance of proper welding jacket, clothing, and gloves for personal safety and protection from sparks, heat, and UV radiation. It emphasizes the need for appropriate attire to minimize the risk of burns or other injuries.

Different types of welders are introduced in the guide, with a focus on MIG (Metal Inert Gas) welders. Recommended models such as Millermatic and Miller Diversion 165 are highlighted, providing an overview of their features and capabilities. The guide also mentions the Miller Dynasty 200 with Coolmate 1 and the Syncrowave 250, showcasing more advanced welding machines.

While the guide provides introductory information on welding equipment, it's essential to receive proper training and guidance from a certified welding instructor or professional. Welding involves various techniques, safety considerations, and expertise that should be learned through hands-on practice and instruction.

By familiarizing yourself with the equipment and necessary items, you will gain a foundational understanding of welding and fabrication. However, always prioritize safety and seek appropriate training to develop your skills and knowledge further in this field.

Contents

Equipment and Necessary Items.

Welding in the off-road fabrication industry is different than welding in general. It will require some specific items and equipment to make your life easier and weld quality better. The first step to weld perfection is being comfortable. If you are stressed or in a very unpleasant position or physical state, then you will likely struggle to produce quality welds. If you talk with all the pros, you will find one very common theme: welding in natural positions with the right equipment.

I know that as you read through the following you will start to see a theme arise – high quality is going to be expensive. But this more than just having the "nice" stuff for the job; it's about your future. Welding can be very harmful to your body if the proper precautions are not taken. Choices like the quality of the helmet and gloves you select will have a direct effect on your future. With protection and injury prevention, please think about your eyesight and other elements before selecting cheap or inexpensive product for welding. Quality components will provide far better results.

With that said, I have had well over a decade of experience in welding for the off-road industry and I can assure you that the products I recommend have been put to the test personally by me in the harshest of fabrication environments. If you are serious about becoming a pro in the fabrication industry, then you will start by obtaining the products that the pros use.

The Welding Helmet!

Disclaimer: Before we start, you need to know one thing: NOT all welding helmets are created equal! There are MANY manufactures out there, but I will only suggest one, with the option of two different styles below. That is for a reason. The helmets suggested below are ANSI Z87.1-2003 (ANSI Z87+) certified. This means that they have undergone rigorous independent laboratory testing to ensure 100% ultraviolet / infrared filtering, high-velocity impact testing, and designed to operate effectively between temperatures of 23 – 131 degrees F.

We will start with what I consider to be the most important element to welding -- the helmet. The reason for the welding helmet is to protect your eyes from the immense bright light and sparks of welding.

Even though you are reading this, I think it's fair to assume that you know NOT to weld without the proper equipment, but I will state it anyway. Under NO circumstances should you ever weld or tack weld without looking though the proper darkness lens attached to the proper helmet.

If you watch some fabrication-type TV shows and hang around some of the pro shops, you might see welders tack-welding without a helmet. Despite how cool they think they look, it is VERY bad for you. I fell under this "cool factor" for a few years. And even though I closed my eyes and tack-welded tubes, plate and material in place, I did some irreversible damage to my eyes. To this day I have trouble with sunlight, computer screens and some lights for even short periods of time like an hour or so.

Learn from my experience. NEVER tack weld without the proper welding helmet, regardless of how "lame" other fabricators will think you are. At the end of the day you will have the last laugh by being able to see better than them well into your later years. And the welding helmet does more than just protect the eyes; it will also shield your face from sparks and excessive UV rays. If you are not wearing a full-face helmet, you will get a very bad sunburn. Skin cancer is NOT cool despite what anyone says. Wear a full-face helmet. Don't be another idiot.

Welding Helmet Lens Selection

There are some misconceptions about the lenses that are currently on the market and how they are applied to your application. Welding-helmet lenses range from #3 shades up to #13 in darkness. If you are welding MIG or TIG, I strongly suggest a #11 to #13 for everything. Some charts and "experts" will recommend as low as #9, but I personally feel as though that is more of a risk than I want to take.

The darker the lenses are, the less damage to your eyes. If it feels like you have sand in your eyes after a day of welding (also known as "flash-burn"), then you have had too much exposure to the bright light of the welding arc. You need to step the shade up and try again. Of course, the darker the shade, the harder it will be to see the weld. That is why many will step down the shade to achieve better weld-puddle vision. You need to find a happy medium. You want it to be light enough to see the puddle properly, but not too light so that damage is caused. As stated above, I prefer #11 to #13 for all off-road fabrication.

It's true that you want to avoid is using too dark of a lens. If you find that you have to squint all day long to even see the puddle, then you need to adjust the lens shade so that it becomes more comfortable for you. If you are having trouble finding the proper lenses, talk to your eye doctor, NOT the guys at the welding store. Reason for this is because everyone is different: a #10 might work for you, but it will not work for me. My eye doctor set me up with #11 and #13. The guys at the welding store are NOT eye professionals; they are salesmen. Do not trust your eyesight to them, or you will end up without it! Anything less than #9 is considered to be a "grinding" lens and should be used for that purpose only. You can get "grinding" lenses as low as #3 for prepping the material and cleaning up welds / tube work. NEVER use a #3-#8 lenses for welding. NEVER!

One major tip I have is that you should properly light the work area. Having proper external light over the work piece will also help with using darker lenses and seeing the weld puddle.

There are two major types of helmets available on the market. Traditional fixed lenses (Passive) and auto-darkening lenses. The traditional type of welding helmet will employ fixed lenses with a specific rating on the darkness. These usually average from #9 to #11, depending on the welder. These helmets do not ever change lens shades unless you physically swap them out. They are dark 24-7, so this will require you to physically flip down the helmet (snap of the neck motion) for each weld. This is because the passive helmet is located in the Up position while the electrode (gun or torch) is positioned. Some prefer this, but many pro welders get tired of it quickly. It makes the start and stop of each weld more difficult because you cannot see (if you keep passive helmet in "down" position)until you create the arc with either the MIG trigger or TIG pedal. When you are forced to tack weld constantly, the Passive helmet creates an inefficient work process because you have to move it from Up to Down position between each tack weld. Not to mention that you will also wear out your neck and possibly create other issues from the repetitive task of snapping the neck to move your helmet down. And if you do not let the helmet lock into its fully down position before striking the arc, you can cause arc flashes, which will damage your eyes.

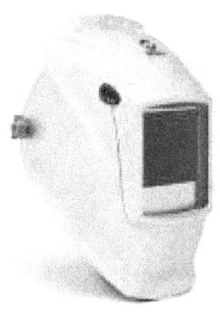

These helmets look like the one pictured above. They are very plain and can be found at just about any welding supply store. They are also relatively inexpensive and typically come with #10 shade, though I suggest that you get a few different shades up to #13 just in case. I personally do NOT recommend these passive helmets.But if you are going this route, then try the Miller "titanium series welding helmet – titanium 1600" passive #10 shade helmet. This helmet features the certification I mentioned above, along with a great viewing area for professional quality welds! But here is the real kicker -- this helmet is upgradeable. Yes, I said upgradeable. You can swap out the front shield flip-down for an auto-darkening unit in the future when you are ready to upgrade. This means you will be able to keep this helmet for years to come!

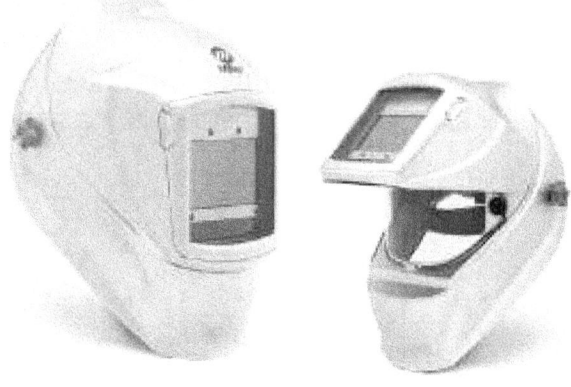

The other major style of helmet is the auto-darkening helmet. This style of helmet provides a special lens that is activated with increases in UV light (when you arc the gun or torch). In other words, it will turn dark when you start your weld and then return to a light shade of about #3 - #4 when not activated, so that you can see your work piece. With this ability to change shade instantly, the helmet can stay in the "down" position before, during, and after your weld. This eliminates the dreaded neck-snapping motion and drastically improves weld quality, because you are more comfortable during the weld process. When it comes to reducing arc flash, these industrial-grade auto-darkening helmets typically will change shade in 1/12,000 to 1/20,000 of a second. Typically the faster, the better. Quality helmets will really make a difference with this. If you are going to be welding all day (start and stop), then you and your eye doctor will appreciate quality auto-darkening helmets. The helmet that I suggest and personally use is the Miller "titanium 9400i" series auto-dark helmet. It has a large viewing window, adjustable shade settings, variable sensitivity (the speed at which it changes from light to dark), and phenomenal grinding features. To top that off, Miller provides you with a 3-year warranty; that is how confident they are in this product.

Proof...

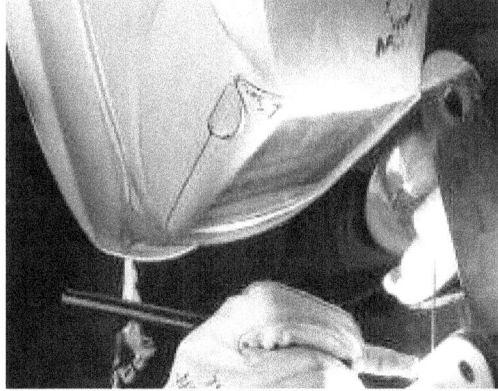

This is a shot of me welding on some caliper mounts on the GMR9 Full Floater rear end for the car to the right. Notice the Welding Helmet and gloves (I elaborate on these next) that I'm using. I not only recommend these products but I use them daily with my work!

Keeping your Helmet Clean

Keeping the viewing area of your welding helmet clean is critical to the success of your welding. You will notice that as you continue to weld, the lens of the helmet will become cloudy and hard to view through. You need to clean it and polish it down to keep it from getting worse or even permanently damaged. I suggest the polish below. Combined with a clean quality rag, it will do the trick.

If your welding lens ever gets damage from sparks or molten metal, then you need to replace it. A damaged exterior lens can cause harm to the internal components of the helmet and those are the expensive parts. You can purchase packs of external replacement lens cheaply, so don't be afraid to replace them when they are worn.

Welding Jacket / Clothing and Gloves

Although not as important as the welding helmet, these items are still very important to the operation of welding and your physical safety. There are a few key elements to be aware of when purchasing these items. Again like everything else in life, you will get what you pay for.

The welding jackets are slightly different for MIG and TIG welding. The MIG welding jacket is designed to prevent burns from sparks and hot material leaving the work piece. I only recommend a cowhide jacket like the one pictured.

The key elements to notice are the material, style, and neck. The material needs to be thick quality cowhide. The jacket I prefer and useon a daily basis is the Black Stallion Standard Jacket. The style is also important. Some jackets only provide leather coverage on the arms and not the body; this is only OK at best. The off-road industry does not have a consistent welding position. You will literally be moving all over the place. If you were to wear a production table-style weld jacket like the one below, chances are that when welding cage work or other elements, you will burn through the center / back portions and your life as a welder will be quite uncomfortable.

The third element to look for is the neck. You want a jacket that is lined around the neck with proper coverage around the neck to prevent sparks from entering the jacket. Notice the ongoing theme about being comfortable while welding? This is for a reason: the right jacket can make welding up a cage / chassis either a breeze or a complete nightmare.

MIG welding also provides challenges with the other clothing you select. Pants and shoes will play a large part in keeping you un-burned and comfortable throughout the welding process. I have welded up quite a few chassis / cages in vehicles, and I can assure you that the proper jeans and boots will save you. Often you will find yourself welding while sitting down Indian style. This means that weld splatter and molten metal can and will travel down toward your legs and crotch. Thick work jeans are preferred, not Dickies work pants. Anything less than nice thick jeans will burn right through. Evenworse they can possibly catch on fire causing real bodily harm. To prevent any accidents to your feet with welding and general fabrication work, I strongly suggest the shoes shown here, CAT high-top steel-toe boots.

They are weld splatter resistance because of the high-quality leather while providing great ankle support. More often than I would like to admit, I have found myself very thankful that I was wearing steel toe boots while working in the off-road industry. It will literally save your feet or toes more than you would think -- parts are heavy!

Here is a little welding tip. Like I mentioned above, some situations will yield unfavorable welding positions with your crotch being in the line of fire, so to speak. I often would keep a little welding blanket handy like this one and cover my lap with it. This will further prevent any un-wanted burns to the crotch while you weld. More importantly it will allow you to focus on the weld and not on your crotch catching on fire.

TIG welding provides a very different atmosphere than MIG welding. With a lack of sparks and emphasis on the freedom of movement, I recommend completely different jackets and clothing for TIG welding. The lack of sparks does not require a full leather jacket like I suggested above for MIG, but instead simply requires coverage. The issue here is the heat and UV rays from the welding. If you have bare skin exposed, then it will be exposed to very harmful UV rays far worse than the sun. I'm not a big fan of skin cancer so I always cover up with a long sleeve shirt that has proper neck coverage. These shirts can be Dickies in form or even a nice cotton long sleeve. As long as you are covered you can wear just about anything, just be sure to not set your arms down or touch hot surfaces.

Unlike MIG welding, you can get away with different clothing in regards to pants and shoes. I often still wear thick jeans just in case, along with high-top Converse. I prefer the extra movement that Converse provides while operating the foot pedal. When I use my steel toe boots, it feels a little funny with the pedal. It might be all in my mind but I feel more in control of the TIG torch with Converse shoes vs. the boots.

Gloves are the next item to address for proper welding and tack welding. I prefer to switch it up a bit when it comes to gloves used for Tack welding and MIG welding. The gloves below are what I suggest for MIG tack welding, either plate work on a bench or tube work in a chassis. They are "Tillman Cowhide Drivers Gloves"

The reason I suggest these is for both protection and longevity. While MIG tack welding, sparks will fly and for some reason they always tend to land right on your hand… or at least mine. The cowhide leather will prevent the sparks from burning through, unlike gloves that are only partial leather and cotton construction. The other reason why I love these gloves for tack welding is that they actually last. Cotton-backed / constructed gloves will tear and fail after they get burned through. And they will get burned!After you have been fabricating for a few days, you will understand: the gloves do not last unless they are leather. I got sick of buying new gloves because of tack-weld burn-through so I bought one pair of those above, and they lasted about 10 times longer than my other go-to Harbor freight Blue gloves.

When it comes to MIG welding (not tack welding parts together), you will need the long, thick, all-leather gloves like these pictured here.

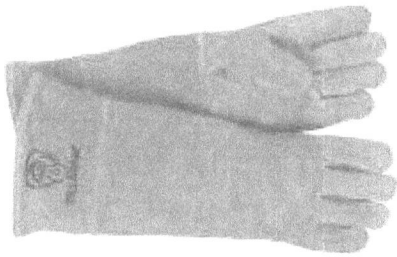

I suggest the all-leather long Tillman gloves; they will cover your wrists and are very durable even with constant MIG welding. They have no problem handling the constant sparks of MIG welding and work great with leaning on hot sections of previously welded cage work. Nice gloves will not only last, but provide comfort while continuous welding takes place.

TIG welding requires a different style of glove with the emphasis on skin coverage and maneuverability. TIG welding is more of an "Art," and you will need all the movement you can get to perform at your best with TIG welding.

This is why I suggest the gloves above. They will allow for coverage up your wrists even with a long-sleeve shirt tucked under the gloves. You do not want the short gloves that only cover up to your wrist. You will end up with 1" burnt strips on the back and inside of your wrists. Again, skin cancer is NOT cool. You want your gloves to be completely made of leather to combat the heat, but thinner than the leather MIG gloves above. The white gloves above are "Black Stallion 25G Goatskin TIG Gloves," and they are my preferred / suggested glove.

Recommended Welders

If you are new to welding and in the market for a new welder, then this section is for you. There are a handful of companies that manufacture welders, but only a few are what I would consider professional grade. A good welder will last you years, and possibly well over 10 years if you take care of it. A cheap welder will not only provide poor results, but they tend to break down right when you need them most. I have well over a decade of experience, and have had employees who literally physically abused my welders... good thing I was never cheap when it came to purchasing a welder.

The two main brands that I suggest are Lincoln or Miller. I'm only going to review the Miller welders below, but for each one there is a comparable Lincoln that is in the same price range. I prefer Miller welders myself; this could be that they are better quality or the fact that I really dislike the color Red... I grew up in a UCLA Bruins family with season football tickets so naturally I was raised to hate everything USC-related, like the color red. With that said, as I type this, UCLA was the victor in last season's rivalry game, so I'm feeling really good about the blue right now!

MIG welders

When it comes to the off-road industry, you will basically only see three different MIG welders. These are the typical welders that are purchased / used by the top fabrication shops across the country. They are in three very distinctive classes, and as stated above you can find a Lincoln equivalent for each of the three below.

For the home fabricator and someone who is just starting out with welding, I strongly suggest the Millermatic 140 with Auto-Set. This little welder packs a punch and is very userfriendly. The new Auto-Set feature from Miller allows you to simply select the wire diameter you are welding with and the thickness of the material being welded. The machine will predetermine the amperage and wire speed for optimal results. This will really speed up the learning curve for those who are new to welding. It allows the novice welder to focus on the technique more than the wire speed / amperage relationship.

The best part about these Miller welders is that it comes with everything you need except a bottle. So you will only need to purchase a bottle, and you are ready to weld. Don't be fooled by the little size of this machine. It can weld up to 3/16" steel, which is more than enough for 99% of off-road fabrication and perfect for the garage builder. The only major downsides to this welder are the small spool size (how much welding wire it can hold), duty cycle, and the lack of a welding cart. This machine can only hold an 8" spool, which can limit you. This means you will have to change out spools more often than the larger capacity units below.

Definition of Duty Cycle –One way of classifying the size of a welder is by the duty cycle rating. This is a universal rating given to welders that determines how hard you can "push" the machine during a 10-minute period. This measure is the number of minutes out of a 10-minute period the welder can operate continuously, with the remainder designated as the "cool down" period. An example of this is a Miller Dynasty 350, which can operate at 300 amps with a 60% duty cycle. That means the machine can operate at 300 amps for 6 minutes with 4 minutes of "cool down" time after continuous welding before starting another weld. If you exceed the welders' limits, then you will overheat the machine and cause irreparable damage to its internals, thus breaking the machine. While the machine is inthe "cool down" stage, be sure to NOT unplug it.

The Duty cycle of this machine is rated for 90 amps at 20%, so for welders it's relatively low. This is typically not much of a problem with off-road fabrication because most welds are less than a few inches, and you tend to take a break in between. With that said, the welder does have a warning light to indicate possible overheating. If you see the light come on, simply let the machine cool down while leaving it turned on and plugged in so the internal fan can operate. All too often I see guys unplug or shut-off the machine when it gets hot. This can cause more damage because it stops the internal fan / airflow that is cooling the critical internal components.

The lack of a cart will prove itself a real "PITA" once you start to weld around a chassis. Being able to easily roll the welder and gas to your desired location is a huge plus, but it's not the end of the world if you can't. It's also aluminum capable, so with the purchase of a spool gun, you can use this machine to weld up to 14ga aluminum. This is a good starter welder. I give it a 3 out of 5.

Millermatic 140

With Auto-Set

If you are more serious about welding and looking to work more than the "weekend warrior," then I recommend you step it up to the Millermatic 212 series with Auto-Set. Just like the above 140, this welder features the Auto-Set system, making life for the beginner very easy. This unit is larger than the above 140 with a capacity up to 3/8" steel and large 12" diameter spools. You will also notice that it's built on wheels. What you can't see in the picture below is that the welder comes with a bottle carrier in the back. It will hold a large (300) size bottle in one simple rolling package, making it perfect for fabrication shops and home builders. The duty cycle is rated at 160 amps for 60%, which is a huge advantage over the 140 miller. Everything you could want to weld while building off-road vehicles can be done with this welder. It's smooth, powerful, and user-friendly. With that in mind it's easy to see why this unit is double the price of the 140. But if you are serious, I strongly recommend this unit. It will last you a lifetime, so I give this welder a 5 out of 5 for the beginner / advanced off-road fabrication welder.

Millermatic 212

With Auto-Set

The third and final welder I recommend is the Millermatic 252. This machine a powerhouse. It's my personal favorite and the one I use on a daily basis. This is a great welder for the fabrication shop and light production facility. You can literally weld chassis all day long 7 days a week, and it will not break a sweat. It's reliable and powerful enough for ½" thick steel material. I have yet to even come close to maxing out this machine, even welding up GMR9 housings all day long!

The down side for this machine is that it does not come with the Auto-Set feature. It can only be setup with manual wire/ amperage settings. But don't be alarmed. The machine comes with a great reference guide to get you started… notice I said *started*. Unlike the top two machined it will only get you in the ballpark. You will have to dial it in with small increases / decreases to the wire speed and amperage. This makes this welder more of a challenge to operate for beginners. But once you get it dialed into your setting, it's a breeze!

Just like the 212, this unit can hold a large bottle and comes ready to roll around your fabrication shop. The increased duty cycle over the 212 is also nice; it's rated for 200 amps at 60%, beyond what any off-road shop will do. I weldhousings all day long with this machine, and those feature 3/8" faceplates and ¼" wall axle tubes, arguably the thickest parts on an offroad vehicle. The large gas bottle and 12" spool capacity will provide hours of welding without changing bottles or spools… *actual welding*, as in with the trigger down. So in reality it can be up to a few weeks of fabrication / chassis welding. Since I'm rating this welder for the beginner, I will have to give it 4 out of 5. The lack of the Auto-Set feature, combined with the fact that you will not be welding fabricated rear end housings all day long, really brought it down a notch compared to the 212.

Millermatic 252

Final Thoughts on MIG machines – *Seriously think about your passion and involvement with welding before you purchase. Saving for the nicer machine will make the difference. But at the same time wasting money on a machine you really do not need is foolish. I recommend the Millermatic 212 for the beginner off-road welder. Ii will last you a LIFETIME and is capable of*

everything you need to weld for off-road fabrication. Anything more than the 212 is like using a bread sword as a butter knife!

MIG Guns and Components

The MIG welding gun is very simple and straightforward compared to the TIG torch. You only have a few components. And typically the consumables last longer with MIG welding vs. TIG welding. There are literally over a hundred different MIG guns and many manufacturers so I can't cover all of them. But I will cover the generic style that you will see throughout the offroad welding industry.

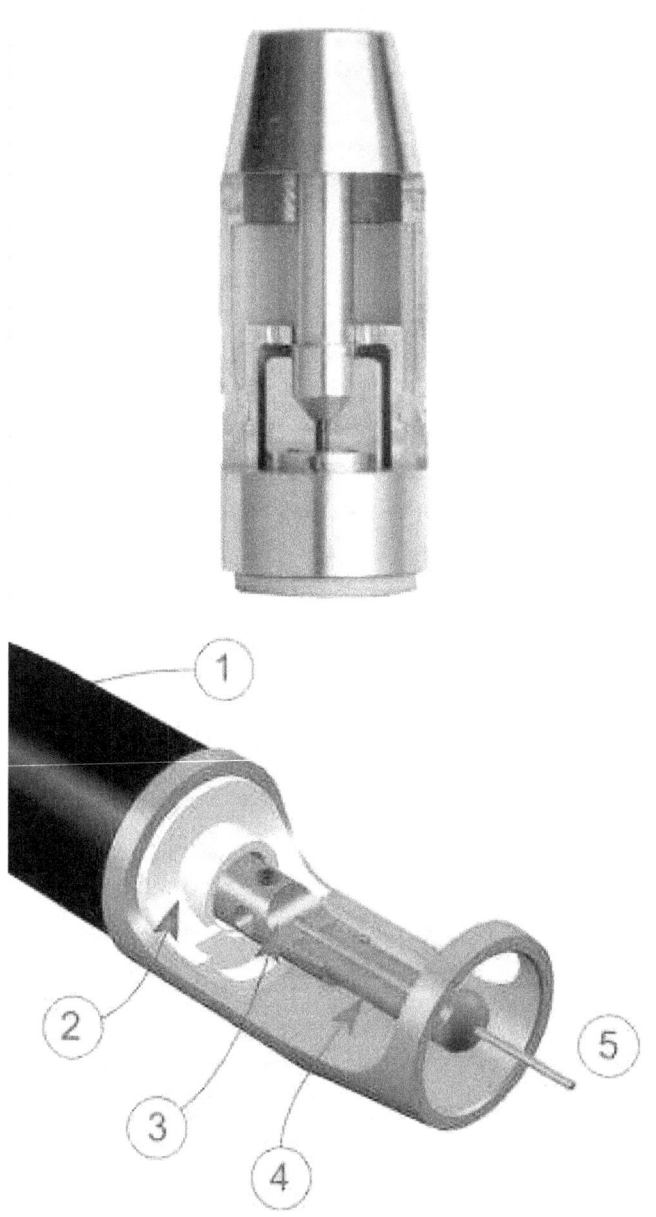

1. **Torch Handle** – This is the main structure of the MIG gun. It holds the end consumables

along with the trigger and internal liner.

2. **Insert** – Plastic molded insert with a metal threaded section (yellow) that accepts the consumables in the torch. Typically this is moldedinto the handle and is not removable.

3. **Diffuser** – This is the part that holds the contact tip and dispenses the shielding gas around the contact tip. Notice the holes in the side, which is where the shielding gas exits the torch handle and travels around the contact tip and inside the nozzle.

4. **Contact Tip** – This is the consumable that has direct connection with the welding wire; it threads into the Diffuser and guides the welding wire to the weld puddle. These tips are typically the first component to get damaged and replaced, so it's a good idea to keep several handy at all times. Another thing to note is that these are specific to your welding wire size. Since they are responsible for transferring current to the wire, they need to be a tight fit. Each wire size has specific tip sizes, so a 0.030" wire will use a 0.030 tip and so on. To ensure proper welding, be sure to use the correct tip for the welding wire.

5. **Nozzle** – The nozzle is the overall cover for the components and the shielding gas director. It protects the consumables and directs the shielding gas to the weld puddle; these can get damaged, especially while learning. If you get the gun too close to the weld, you will melt the ends of the nozzle into the weld puddle. Then you will have to replace it, so it's a good idea to keep a few of these handy as well.

Trimming the MIG wire – Before you start welding, you need to make sure you have the optimal wire length to ensure a proper weld bead start. The proper length is going to be about 3/8" to ½" in length past the end of the gun, as pictured below. If you do not start with this, then you can cause the welder to have trouble starting an arc or even delay the arc. The optimal length will make starting easier to control so you will have an aesthetically pleasing weld.

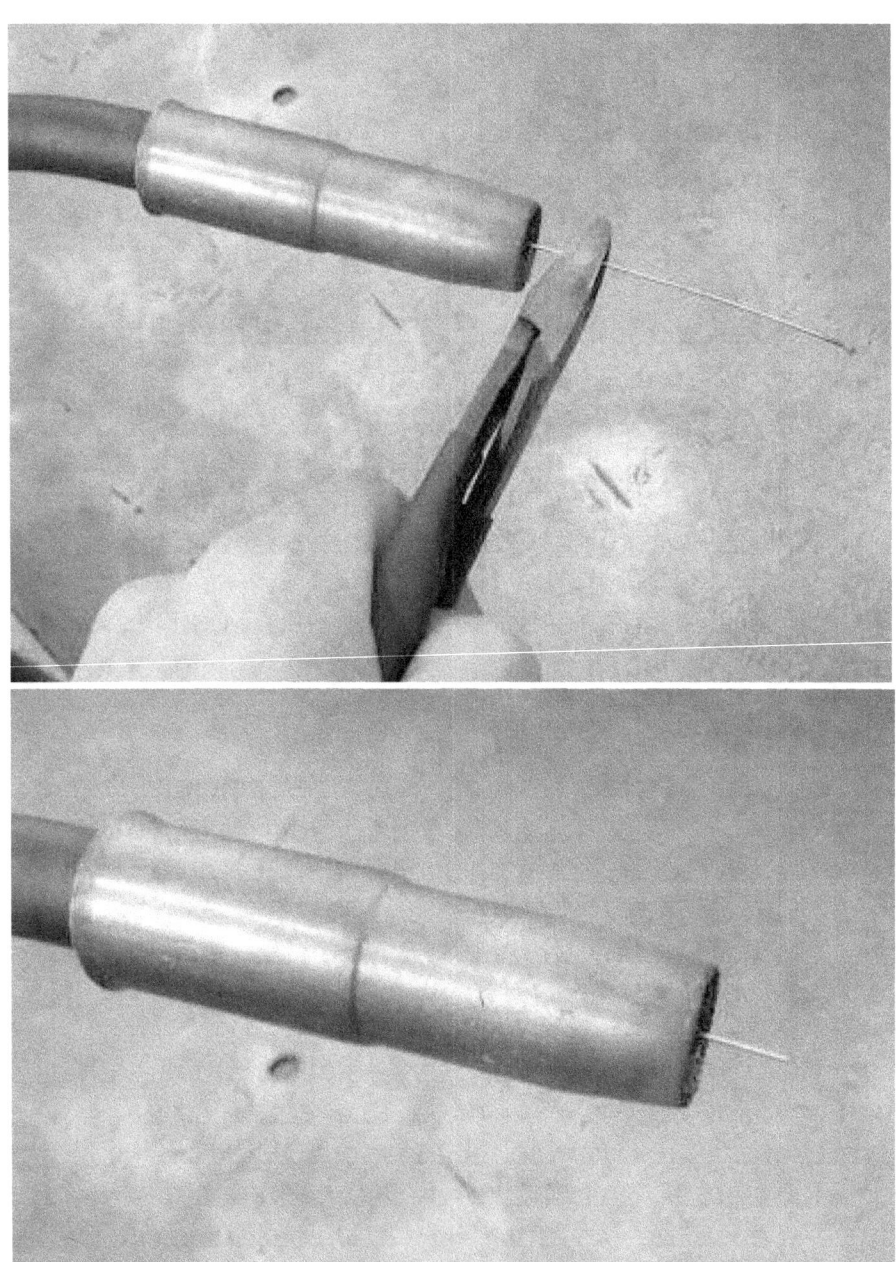

Nozzle Dip – One of the most useful tricks to MIG welding is the addition of nozzle dip to your welding process. You can find this at any welding supply store, and I suggest that you pick one up next time you are there. The process is simple: First, you need to get the end of the gun warm, so weld on a scrap piece of metal for about an inch. Right after you are done welding, clean the inside of the nozzle and weldingtip immediately. Be sure to do this quickly so the nozzle does not cool down. Otherwise the dip paste will not properly melt into the nozzle. Then dip the nozzle of the gun about 5/8" into the paste. At this point you should see the paste melt to the nozzle and accumulate around the inside. You will end up with excess dip on and in the nozzle. Now, with the same scrap piece of metal, weld another inch or so to melt off any excess paste from the interior of the gun. While the gun is still warm, take a rag and wipe down the exterior of the gun. You are now ready to weld your work piece. Repeat this process about every 30 minutes. To give you some reference, I do so for every two spring hangers I weld up or about twice per rear GMR9 housing.

 **TIG Welders**

Just like the MIG welders that I listed and recommended, I will only focus on the Millermanufactured components, in my experience they are the best quality / performance. If you are more of a Lincoln fan, there are comparable Lincoln welders that match the below units within reason. I will start off small, then work upto the big boy of the off-road fabrication industry. There are two major differences among the TIG welders that I will recommend: the first is considered to be "air-cooled," and the latter two are "water-cooled" units. Preventing overheating of the torch/ lines is critical and must always be monitored.

For those new to TIG welding and working on projects out of your garage, I suggest the Miller Diversion 165 TIG machine. Just like the above smaller MIG machine, I recommend this setup for the home shop beginner, and it comes with everything you need to get the ball rolling except the bottle. All you need is a bottle, and you are off to the races, literally. This little unit provides a very simple setup-and-go system. All you have to do is power up, select material type, and then set the thickness range. It takes the guess-work out of setting up the machine. The power (arc) of the welder will be fine-tuned with the foot pedal as you press down. The machine automatically gets the proper range for heat. All you have to do is focus on the technique. Less worry means more focus and improved learning curve.

This setup is called an "air-cooled" setup because the torch and lines are only cooled by the surrounding air. This is great for beginners who are not welding for long periods of time or with very high amperage. Being that it is air-cooled means less setup and cost, so it's cheaper to get started with the TIG welding game. This unit also features the ability to weld up to 3/16" steel and aluminum, with a duty cycle of 150 amps at 20%. Being that it is air-cooled and has a low duty cycle means that you will spend more time waiting for the lines / machine to cool than actually welding. With that being said, this machine can TIG weld almost everything you will see when welding in the off-road fabrication industry. Despite the small size and air-cooled setup, I give this machine 3 out of 5.

Miller Diversion 165

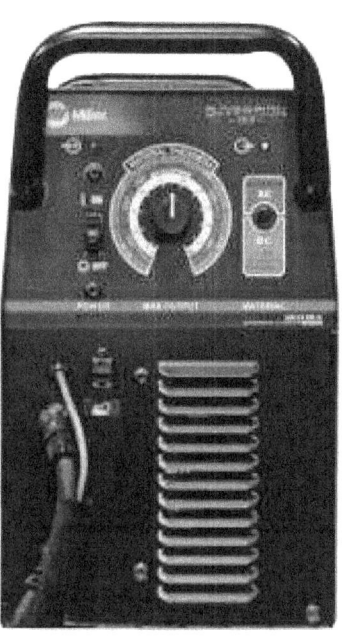

BONUS – The Miller Diversion 165 can become a water-cooled machine. All you will need is to purchase a 1-gallon water cooler and new lines that will support the water-cooled torch. Contact your welding supply store, and they can get everything setup for you. The downside to this upgrade is that you will be spending just as much as on the setup below once you get a new cart and larger bottle, thus only making this a3.5 out of 5 with the water-cooled setup. If you are planning on going this route in the near future, then you should already be looking at the TIG machine below.

The next two welders I recommend are "water cooled" units. If you look at the diagram below, you will see the layout of the water-cooled welder with the machine, coolant system, and the gas bottle. You will also notice that the water is referred to as "coolant." This is because it is not actually water. It's a coolant that Miller has designed specifically for TIG welding conditions and systems. The reason I refer to it as "water cooled" is because that is what the old timers who taught me to weld called them. To this day, some of the welding supply stores still refer these TIG machine systems as "water cooled" as well. So to be clear: Coolant or water-cooled TIG welding machines / systems are the same thing.

Professional Fabricator Tip – When purchasing your TIG machine with water cooler, be sure to get the nicest lines you can. The cheap rubber ones that typically come standard with all welders are CRAP! They will break, crack, leak, and cause you some very frustrating days! Get the quality lines like the ones pictured below with a very tough external sleeve, something like Kevlar or leather. Protecting these lines is critical. If you break one or get even the slightest leak, you are instantly done welding for the day until you replace the entire line!

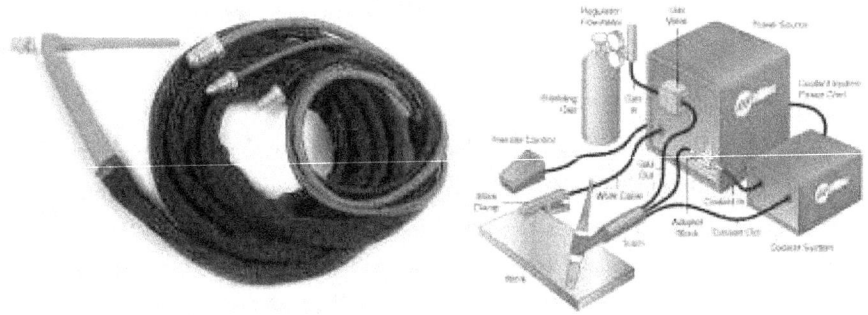

If you are more serious than the "weekend warrior" welder and looking for something that can do it all, then the Miller Dynasty 200 with Coolmate 1 and welding cart is for you. This setup pictured below (complete on the right) is a complete water-cooled setup. This welder is capable of steel / aluminum thicknesses of ¼" whichis perfect for off-road fabrication. As you will see later on, I prefer to TIG weld slightly lower amperage than suggested with many parts, so in reality you can weld 3/8" with this. I have and do regularly weld up fabricated rear end housings with 3/8" face plates and ¼" tubes all day long. Even in 120-degree Arizona weather, this welder and cooler can keep up with a full dayof welding abuse. If I can't max out this welder, than you definitely will not! It is rated for 150 amps at 60% duty cycle and 100 amps at 100% duty cycle. That means you can weld .095 and .120 plate / tube all day long without stopping! Pretty much 90% of the thicknesses you will see in off-road fabrication.

The setup is a little more difficult than the above 165, but with some practice you will have your settings down in no time! One major advantage of this setup is how compact it is. The complete setup with a large welding bottle takes up the same amount of space as the above Miller 212 and 252 welders! Its lightweight / compact stance allows for easy use around the shop and chassis fabrication. One element that is a MUST with the setup is a 25' high quality torch / cables (like stated above) setup that will allow you to climb through a chassis and weld every nook and cranny! This setup is the Ideal TIG welding platform for off-road, period. I give this package 5 out of 5.

Miller Dynasty 200

With Coolmate 1

The third and final TIG welding setup I recommend is the Miller Syncrowave 250 with the matching Coolmate 3X cart. This is the setup for professional shops. I have owned two of these welders in the past and have nothing but great performance experiences with it. Everything I threw at it was not even close to maxing out this machine; I literally could not get it to come close to overheating. It is capable of welding ½" thick steel and 3/8" aluminum so anything off-road related will not be an issue. Just like I mentioned before, I would use this machine to weld up fabricated rear end housings all day long, no issue! The Duty Cycle is rated for 250 amps at 40% and 200 amps at 60%. This machine will outlast the others, with a track record to prove it from Miller. There are a few downsides: this machine is HUGE… literally taking up double the amount of space as the above 200 series welder. I only recommend this setup for the larger shops and more advanced welders. It can be a little tough to deal with around a chassis, but with practice this machine can be dialed in to perform beautifully. Just like I recommend with the 200 series above, you must get the 25' high quality torch and cables. Another advantage of this machine over the 200 is the fact that it can hold two large welding gas bottles, thus making it better for production or daily fabrication operation. The overall size and cost make this welder not the ideal setup for the beginner, so I have to give it 4 out of 5.

Miller

Syncrowave 250

With Coolmate 3X

Final Thoughts on TIG Machines – *The purchase of a New TIG machine is going to be something that will take some thought. Consult your local welding supply store on all the costs involved so that you are aware of what the entire setup will be -- not just the machine. Not all*

off-road fabricators even use a TIG machine. So if you only start off with a MIG machine, do not worry, in time you can work your way into TIG welding. With that said, I strongly recommend that you focus on the most bang for the buck, which will land you on the Dynasty 200 series and Coolmate1 cart. I'm confident that you will experience many years of fabrication and welding with the 200. It's a beast in a small package, and that is why I use it on a daily basis!

Now that I have reviewed the machines that are available, let's take a close look at the control elements, such as the torch and the pedal components. These are the direct tools that you will use to melt metal and create off-road art!

TIG Torches and Pedals (controls)

It is easy to get confused and frustrated when itcomes to the controls of TIG welding -the cable, torches, and pedals all have many options. I will review the basics and some that pertain specifically to the Off-road fabrication industry. There are some I will not cover because they are far too rare and not seen in this style of welding, (they're more often seen with very unique materials in aerospace welding). The overall generic items I point out below can be found in all the common forms of TIG torches that come with the above welders I recommend.

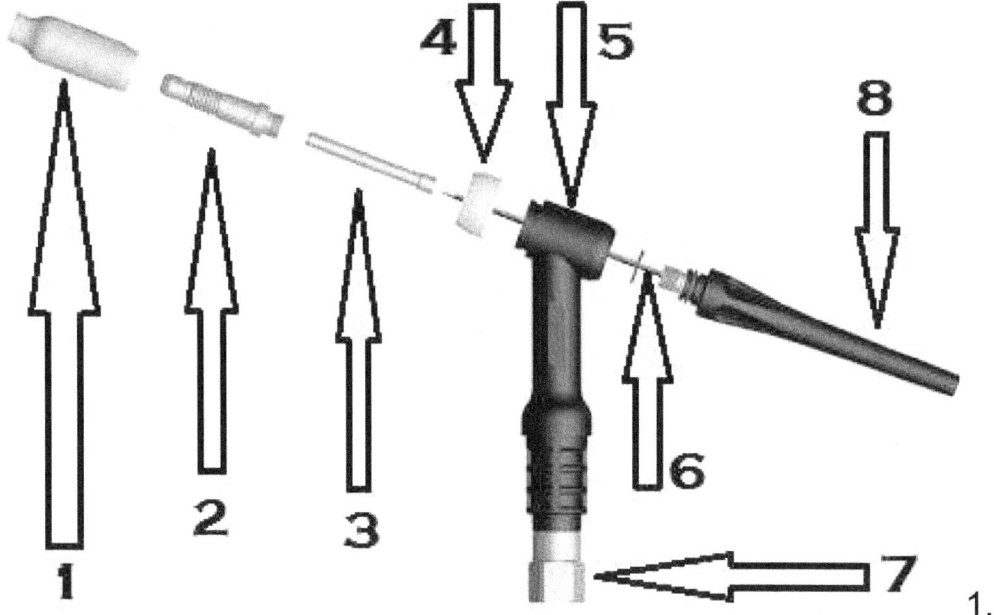

1. *Cup / Nozzle* – Threads over the Collet Body and is used to cover the tungsten and direct the flow of the shielding gas.

2. *Collet Body* – Threads into the torch body and used to hold the Collet and direct flow of shielding gas from torch body to the Cup / nozzle.

3. *Collet* – This piece is what gets sandwiched between the Collet body and the tungsten. The Back Cap holds it in place and ensures a tight fit so that the power can be transferred through the torch head, to the Collet Body, then Collet, then down the tungsten.

4. *Cup Gasket* – Threads between the Collet body, Cup, and the torch body. This is used to seal the unit and prevent any unwanted air in the shielding gas flow stream.

5. *Torch Body* – This is the case that holds everything and connects the power cable, water, and gas lines to everything.

6. *Tungsten* – This is what creates the arc with the material and forms the welding pool. (I explore this in depth below.)

7. **Cable Connections** – This is where the power, water, and gas cables are connected to individually.

8. **Back Cap** – This is what seals the backside of the torch and protects the tungsten from accidently creating an arc with anything.

Back Caps for the TIG torch

The back cap is the end of the torch where you will adjust and tighten the tungsten in the torch body. When you need to replace, sharpen, or even adjust the tungsten, this is the first component you loosen. It is a right-hand thread and just like everything else, right = tight and left = loose. To accommodate the different positions you will be welding in, there are three different tungsten lengths: standard, medium, and button.

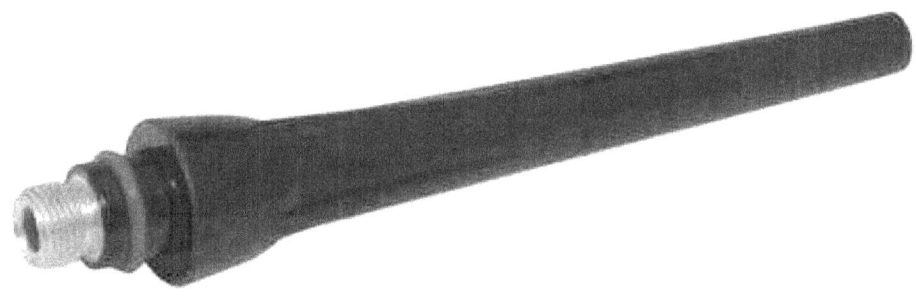

The Standard Length back cap is designed to be used with all off-the-shelf 7" long pieces of tungsten. These are the common lengths and are found at all the welding supply stores. The downside to this (pictured above) is the size. It'seasy to notice that the length of the standard end cap is more than the entire torch – cap assembly. This can cause some problems when welding in off-road related situations. Welding around a chassis or truck will be difficult; not to mention some component welding such as arms can also be difficult, even if on a welding bench. I do not suggest this end cap for your torch even though it comes standard on the welders I mention above.

The medium end cap will accommodate up to a 3" tungsten and provides quite a bit more clearance for the back of the torch. Even though tungsten does come in shorter lengths, I do not recommend buying it short, it's expensive and still hard to find at welding supply stores. I recommend that you still purchase the standard length 7" tungsten and simply cut them down. You will end up with double the pieces of tungsten for less than half the cost.

The third and final one is the "Button cap" and is the smallest of the three. It will only accommodate tungsten in lengths of about 2" but it does provide the most amount of clearance for the back of the torch. I only recommend this end cap for experienced guys; otherwise you will become frustrated quickly.

Part of welding for off-road fabrication is efficiency, and the Button cap is only efficient if you are very experienced with TIG welding. While you are in the learning process, you will undoubtedly mess up the tip of the tungsten quite a bit, thus requiring you to resharpen it, loosing little bits of length each time. The Button cap only allows for you to sharpen the tungsten for about 5/8" before the tungsten will not tighten properly inside the torch head. This means that if you are using a Button cap, then you are much closer to throwing away perfectly good tungsten simply because it's too short.

For that reason I suggest the Medium end cap for the beginner. It will allow you three times more sharpening than the button cap, and that is very beneficial. Once you have mastered the technique and you can weld without messing up the tungsten, then switch over to the button cap. I use the button cap, but still recommend the medium for those starting out.

Types of Tungsten Electrodes for TIG welding

Tungsten is a very common material used in the industrial field. Not only is it used in all TIG machines, but it is also used to make cutting tools such as saws and CNC machine tools. There are 6 main tungsten colors that represent 6 different electrodes that are used for specific welding applications. Do not get overwhelmed by this section. It's simple to understand, and I review what I personally use in the next main section.

GREEN – PURE Tungsten : This tungsten is designed for AC-current welding and primarily used in welding aluminum. It features the ability to "ball" the end of the tungsten easier than others and is also the cheapest tungsten to purchase. With good arc stability and decent resistance to contamination, this is a good tungsten to use in all low-current AC applications.

ORANGE – 2% Ceriated – EWCe-2: With a minimum of 97.3% pure tungsten, it can be used as a replacement to the "Red – Thoriated" electrodes. With low amp arc capability this tungsten will work with both DC and AC applications. It can be used in the place of "Red" and most welders will not even be able to tell the difference. Because it is 97.3% Pure and 1.8-2.2% Cerium, it does not perform great with higher amperages because the oxide content can get burned away.

RED – Thoriated 1-2% - EWTh-1 and EWTh-2: This is the most common Steel and nonferrous metal tungsten with great performance for DC welding of all amperages. The 2% version is the preferred with the best arc start, stability, longevity, and resistance to corrosion. The end result is that the sharpened end of the tungsten will last longer and provide better performance with the 2%, a plus for all welders, both beginner and advanced.

GOLD – Lanthanated – EWLa-1.5: The Gold tungsten is similar to the "Red" and can replace it without major changes to the settings and welding techniques of the machine / welder. It has the same positive attributes as the "Red" and is used for DC current applications.

BROWN – Zirconiated – ERZr-1: The Brown tungsten is great for AC-current welding due to its increased ability to maintain the "balled" electrode end. With superb resistance to contamination, this tungsten is the preferred electrode in welding situations where contamination of the weld is not tolerated. This can be used as a replacement for the "Green" tungsten with welding materials like aluminum.

BLUE – 2% Lanthanated EWLa-2: This tungsten is used to weld Stainless Steel, Inconel, nickel / cobalt alloys, titanium, and most other aerospace alloys. It provides great arc-starting and current-carrying, while maintaining low erosion with 200+ arc starts. Used for DC-current welding and not commonly found in off-road fabrication shops or applications.

Sharpening Tungsten

Although this is critical to the success of your welding, it is not as important as some maintain. Often you will hear guys rant and rave about how you have to have a professional $1000+ tungsten sharpener or you are "doing it wrong." Well, that is just plain BS! I do not use and never had a fancy Tungsten sharpener. I prefer to use either my hand with a 90-degree Air dir grinder / Roloc disc or drill with DA sander. Some of the aerospace welders will flip out that I said that but when it comes to off-road fabrication welding it works more than "just fine." Before I review how I sharpen the Tungsten, let's look at how it's done in a perfect world.

The main issue to avoid is grinding the tungsten in the wrong direction. The key is that you want to get the taper of the end as symmetrical as possible. Remember that the geometry of the taper will affect the arc, but typically it's not very critical in off-road fabrication.

Shown below is the incorrect way to sharpen your tungsten. Notice the direction of the tungsten relative to the grinding wheel. This method makes it very hard to control the symmetry of the taper no matter how steady you hold the tungsten. The other issue is that it tends to create radial marks down the taper of the tungsten and this can add to the unusual geometry and cause the arc to be inconsistent.

INCORRECT

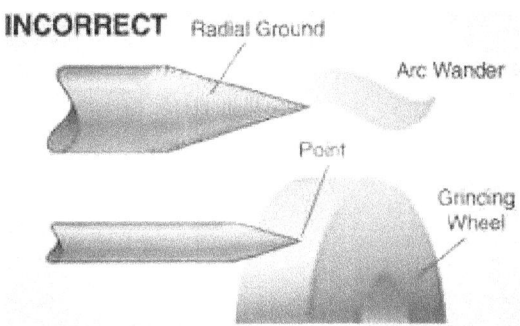

To properly sharpen the tungsten, you want to follow the path of the grinding surface/ wheel as in the picture below. One of the best ways to accomplish this is to utilize tools around the shop that will lend a hand with turning the tungsten while you sharpen it. I prefer to use a cordless electric drill.

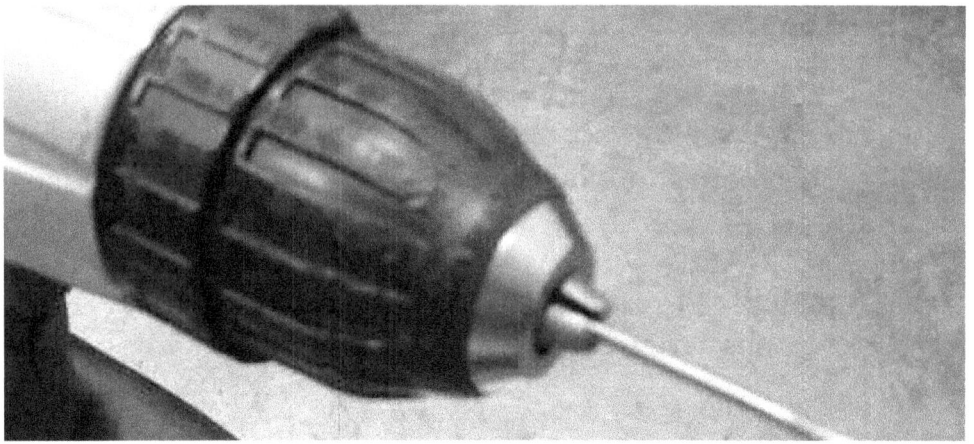

You can also use an air-powered drill to rotate the tungsten while you sharpen it. This method will make it very easy to maintain the consistent geometry of the taper.

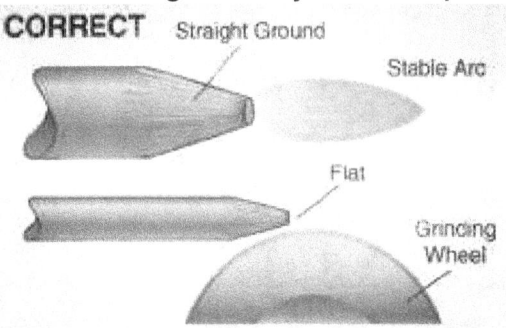

TIP - While sharpening the tungsten, I recommend that you use a specific and separate abrasive so that you do not contaminate the tungsten with other materials. In the picture above, it's easy to see that the wheel is only used to sharpen tungsten. Another tip is that the Sharpness or Bluntness of the electrode will have an effect on the outcome of the arc.

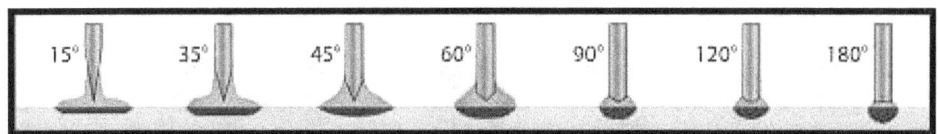

When you utilize sharp tungsten, you will notice that the arc is more stable while welding with fewer amps, which is great for thinner steel materials like Chromoly sheet metal and tube work. As you reduce the angle of the taper (sharper), you will have less penetration, shorter electrode life, wider weld bead, and easier arc starting. When you head to the blunt side of the table, you will notice the arc wandering (higher chance) but can weld with more amps. The Weld penetration and electrode life is increased while the bead is narrow with harder arc starting.

With all that taken into account, I prefer to weld Steel and most of the components / chassis work in off-road with about a 45-35 degreeangle on the electrode.
When it comes to Aluminum, I prefer to be in the 60 range with the balled end.

My preferred method of sharpening the tungsten is with the cordless drill and DA sander using 320-grit. It's fast and easy to use when in the fab shop. A trick I like to use is to sharpen about 3 to 4 tungstens on both sides at once; so that once one is used, I can swap in another without going back to the sharpen mode. It saves time and can prevent frustration headaches when you have a lot of welding to accomplish in one session.

TIG Welding Cups / Nozzles

One of the most noticeable components of the TIG torch is the Cup / nozzle. This end cap is designed to direct the flow of gas and protect the electrode while welding. There are three different types commonly found with TIG welding. To complement the performance of the cup, welders will often prefer to utilize Gas lenses, which I will get to in a minute.

The most common TIG cups are manufactured from "alumina oxide" and have a very distinctive pink color to them. They are the most inexpensive and commonly come standard with all TIG welding packages from Miller / Lincoln. The quick and simple manufacturing process of injection molding keeps the cost of these cups down. Due to the chemical composition of the cups they are often referred toas Alumina Cups. The ability to provide great impact resistance when used in general low amperage TIG welding makes them great for the beginner and everyday fabricator.

If you are using the Alumina cups in very high reflective heat conditions, it is not uncommon that the cups can break. When using this cup in a confined space, the possibility to overheat the cup material is high, causing it to crack when it cools. Once an Alumina cup cracks, it means that it is only a short period of time until it breaks.

I suggest that if you see your Alumina cup start to crack that you replace it. That way you do not risk a cup breaking while you are welding, causing issues with the weld quality. It's easier to replace a cheap cup when needed, than to grind out a weld and rework the area.

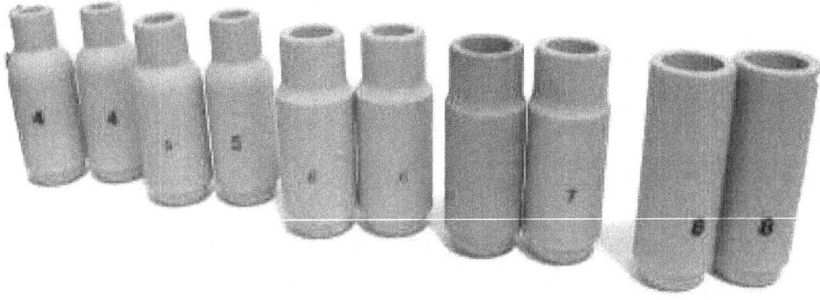

The second most popular cups are more of a tan / grey color and are comprised of a high-temperature non-conductive clay material that is manufactured by machining the cup size on a lathe. Being that this is a different manufacturing process, it lends itself to providing many different shapes and sizes of cups that typically cost more than the above Alumina cups. It is often possible to find XL and even XXL cups made from the "lava" material. These cups work great in specialty applications where high heat / amperage is present but they still suffer from the same reflective heat issue as the Alumina cups. If they are in a confined space, cracking / breaking can occur from the heating / cooling process.

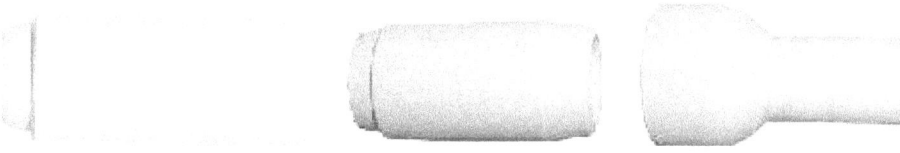

The Third and most expensive welding cups are the Silicon Nitrate series, and to no surprise they are also the highest performing cups. With the ability to stand higher amperage and temperatures, these cups typically last longer than both of the above cups. This longevity provides the cups with the highest duty cycle of all three cups, along with the ability to provide better consistency with higher-quality TIG welds.

These Gas cups consist of two different types: Pyrex and Quartz. The Pyrex cups are lowtemperature nonconductive; while the Quartz is high-temperature nonconductive. Both are manufactured by being "blown," so they are unable to be threaded, thus requiring you to convert your torch to a press-on-style cup / nozzle.

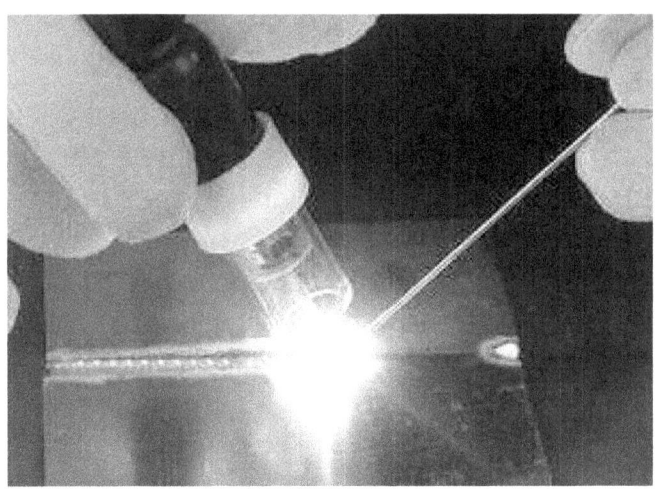

Below is a TIG welding Generic Flow chart for gas and cup selection reference.

GUIDE FOR SHIELD GAS FLOWS, CURRENT SETTINGS AND CUP SELECTION									
Electrode Diameter in inches (mm)	Cup Size	WELDING CURRENT (AMPS) - TUNGSTEN TYPE				ARGON FLOW - FERROUS METALS		ARGON FLOW - ALUMINUM	
		AC Pure	AC Thoriated	DCSP Pure	DCSP Thoriated	Standard Body CFH (L/MIN)	Gas Lens Body CFH (L/MIN)	Standard Body CFH (L/MIN)	Gas Lens Body CFH (L/MIN)
.010 (.25)	4 or 5	5-15	5-20	5-15	5-20	5-8 (2-3)	5-8 (2-3)	5-8 (2-3)	5-8 (2-3)
.040 (1.0)	4 or 5	10-60	15-80	15-70	20-60	5-10 (2-4)	5-8 (2-3)	5-10 (2-4)	5-10 (2-4)
1/16 (1.6)	5 or 6	50-100	70-150	70-150	80-150	7-10 (3-4)	8-10 (3-4)	8-10 (3-4)	7-10 (3-4)
3/32 (2.4)	6 or 7	100-160	140-235	150-220	150-250	10-15 (5-7)	8-10 (3-4)	10-20 (5-10)	10-15 (5-7)
1/8 (3.2)	7 or 10	160-210	150-325	225-330	240-350	12-15 (5-7)	8-12 (4-6)	12-15 (5-7)	10-20 (5-10)
5/32 (4.0)	8 or 10	200-275	300-425	375-475	400-500	15-20 (7-10)	12-15 (5-7)	15-20 (7-10)	10-25 (5-12)
3/16 (4.8)	8 or 10	250-350	400-525	475-800	475-800	20-30 (10-15)	12-25 (6-12)	25-40 (12-18)	15-30 (7-14)
1/4 (6.4)	10	325-700	500-700	50-1200	700-1200	26-50 (12-24)	20-35 (10-17)	36-55 (14-26)	26-45 (12-21)

TIG welding Gas Lenses – Most welders are familiar with the components that I pointed out above: the cup /nozzle, collets, collets bodies, back caps, and tungsten. But the introduction of a "gas lens" will typically confuse them. It is rare, and only few know about them. And even fewer use them. They are often seen being used with high performance products, low access joints, and specialty alloys like Titanium. The increased cost of these components is the main reason why they are not often used. It's more of a "would be nice" item in the off-road fabrication industry than a "need" item.

The Gas Lens itself replaces the Collet Body, and with the Collet it holds the tungsten in place. Typically gas lenses are designed with several mesh screens cut from stainless steel that are housed in a copper body. Some of the cheaper gas lenses will use mild steel mesh, but I don't recommend those. They can rust and corrode easier, which will cause gas flow problems. These lenses are commonly available at any welding supply store for both air and watercooled torches.

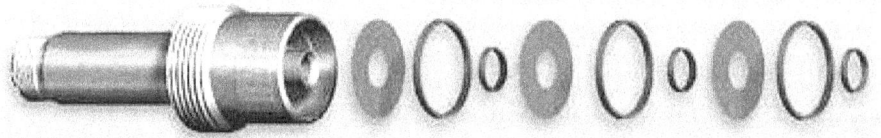 The benefits of a Gas Lens are very clear in the picture below, showcasing the gas flow of a gas lens (left) to that of a traditional cup (right).

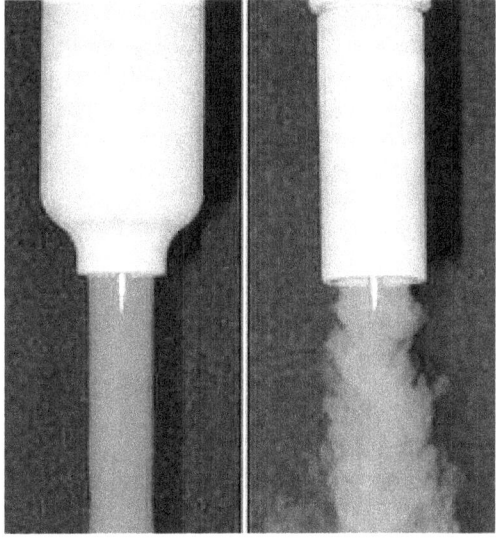

It's clear to see the linear flow path of the shielding gas and why that is a benefit. The flow is not only longer but also entirely undisturbed as it exits the cup, making it very predictable and consistently covering the tungsten.This allows you to move the tungsten farther out of the cup for improved visibility. These lenses are extremely helpful with materials such as Titanium, Stainless, and Aluminum. The Gas lens helps reduce weld porosity and material degeneration with materials that are highly reactive to atmospheric contaminants. Even in basic Steel TIG welding applications, the improved Gas flow will help with shielding gas coverage and consistent welding performance.

With the ability to stick out the tungsten from the cup (in some cases as far as 1"), the gas lens is the ideal setup for roll cage welding. Getting into tight places with low visibility can be made easier with the use of a gas lens. For complex tube junctions that are often seen with off-road, I suggest the use of a gas lens. If used properly, the gas lens can actually use less shielding gas. In some situations, it can lead to 3 to 5 cubic feet per hour less, so in time it can pay for itself.

When you are installing a Gas lens onto your existing torch, it is important to add a transition insulator. The use of a gas lens will require a much larger cup, like this one pictured.

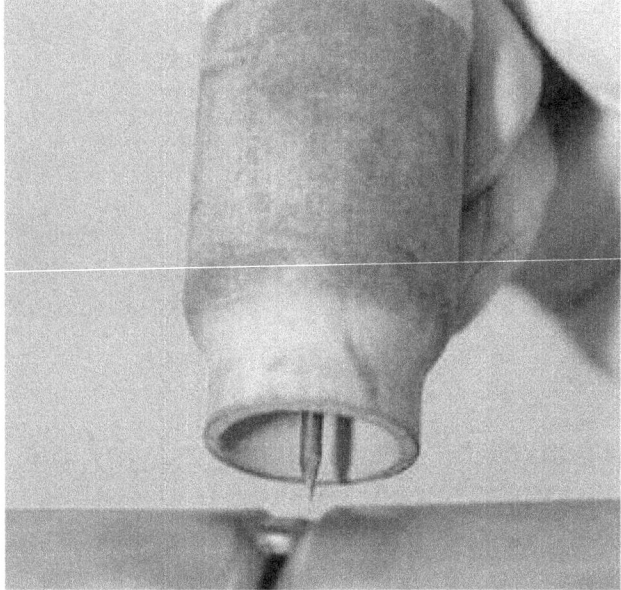

The transition insulator seals the threads where the cup / nozzle screws into the gas lens, a very important component to the success of welding with a gas lens. The insulator prevents air from entering between the cup and thegas lens. This ensures that the flow of the gas out of the cup will not be contaminated. Keeping an eye on the insulator and making sure the install of the cup is perfect will prevent any issues with poor shielding gas quality.

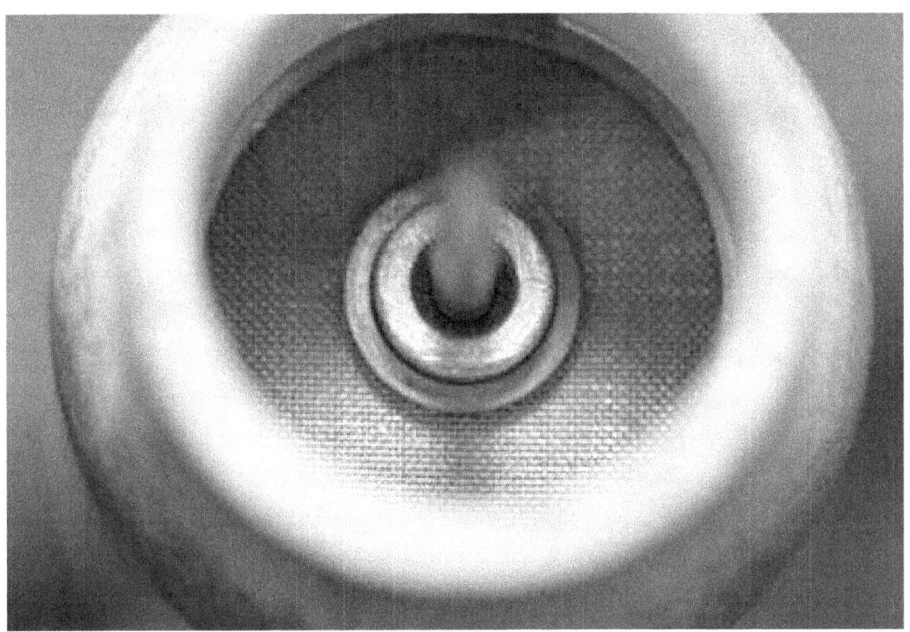

Just like all consumables, the Gas lens and the mesh can become worn / damaged over time. This damage will disrupt the flow of the shielding gas and make the gas lens ineffective. Weld spatter and puddle popping can damage the mesh, so keep an eye on it while you are welding. The picture above is what the gas lens should look like. Visual inspection is the best line of defense for keeping you welding with consistent flow.

Controlling the Amperage (TIG Controls)

TIG welders have advanced to the point that you can pick different methods of controlling the amperage that the machine puts out for different positions. The Typical TIG machine will be set to max amperage, and the control will allow you to start / increase the arc up to the set amperage on the machine (which you set manually). The control at full engagement (either pedal all the way down or thumb control maxed) will max out the amperage set on the machine. For example, when the machine is set to 120 amps and the pedal is completely down, you will be at 120. When the pedal is halfway down, then you are at about 60 amps.

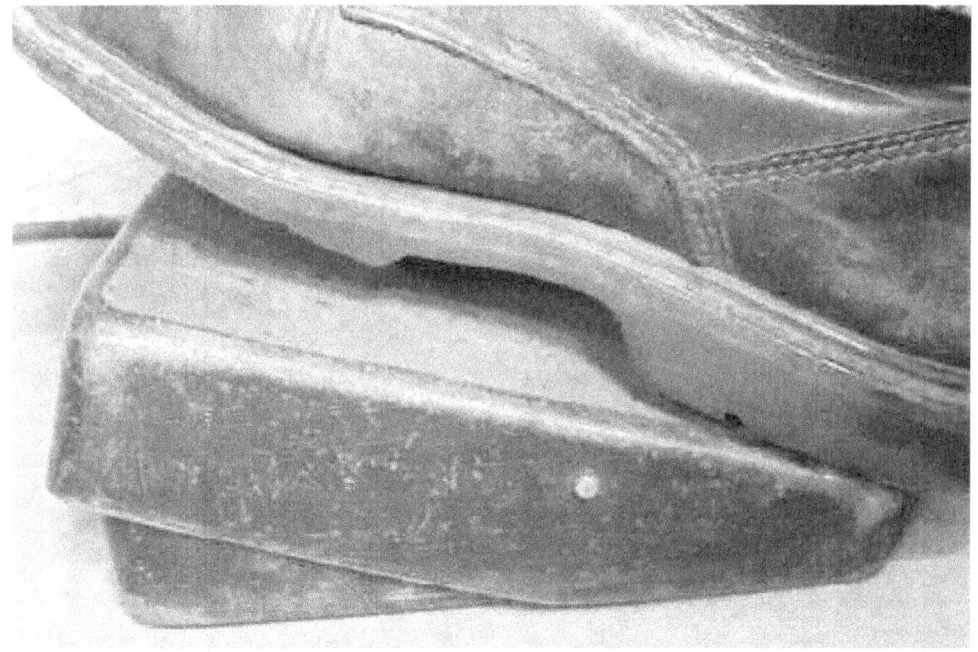

The most common and standard control (with most TIG packages) is the traditional corded foot pedal. Using it is the best way to learn. It's effective and will teach you control. I suggest that you take the time to learn to controlthis with both feet, and between your knees. When welding in a chassis or in cage work, you willbe forced to control the pedal with other parts of your body, such as your knees.

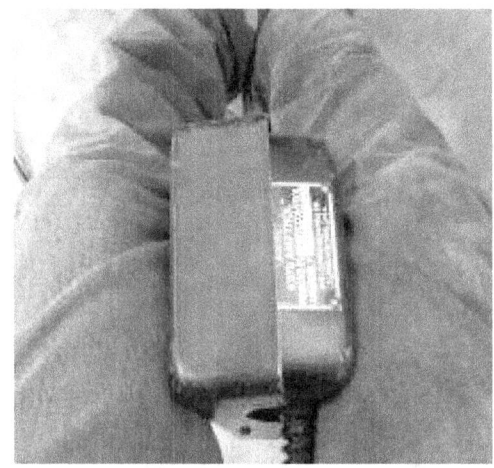

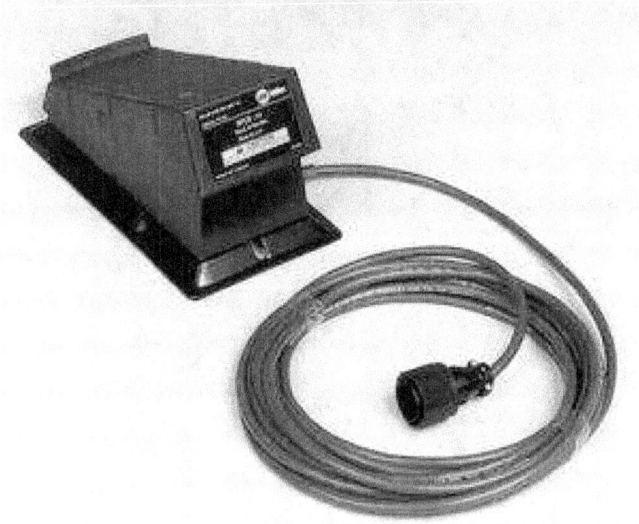

With improvements in technology we have see the progression of the wireless foot pedal. These are great and I strongly recommend them, but the downside is the cost. They are well over double the cost of the above unit. Of course, the best part about them is that you will not have to deal with the cord, which can become very annoying when chassis welding.

If the foot pedal assembly is not working out for you, you do have other options. You can control the amperage with your thumb, located on the torch. You will find these in some high-end fabrication shops. They are designed for areas that are hard to reach and can provide easier access and welding for chassis situations. The downside is that you do have to control it with your thumb; I do not prefer this method. I actually find it more challenging and harder to focus, because you have one hand controlling the torch and amperage all at once.

There are two basic styles of this control: the first is the removable kind, as pictured below, that will attach to the torch. I do not recommend these. They typically do not last, and they will move around on you, making the welding procedure far more difficult.

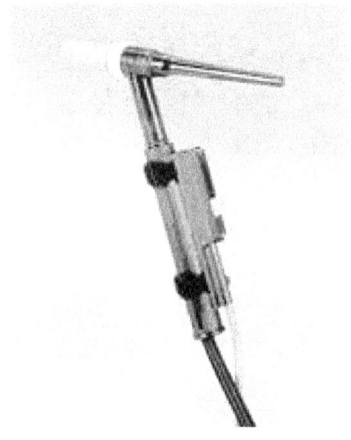

The second kind of hand control is one that has the thumb controls built into the torch like the picture below. They are expensive but arewell worth the extra money if you are serious about using a hand control. This torch combined with a button cap will provide a great setup that can be used in very tight situations like tube chassis welding.

Even with all the advancements in the controls that can be used with TIG machines, I prefer the standard corded foot pedal. They are reliable, easy to use, and are still used by countless top industry fabricators.

MIG welding Wire – With all the know-it-alls in the off-road industry, it's easy to get confused about what welding filler to use and for what. Keeping with the "Cut the Crap" theme, I will make this very simple for you.

ER70s-6 : This welding wire can be found in all sizes / spool diameters and is the most economical of the welding wires. Be sure you get the "-6" filler; the others are not going to be optimal for what you will be welding in off-road fabrication. This is a good wire, but not great.

ER80s-D2 : The optimal wire for everything off-road. Very few utilize this wire, but it's quality is better than the ER70s-6 and can be found everywhere for only a small amount more than the above.

Stainless Steel : This is a little tricky. The welding wire will reflect the base material you are welding; for this you need to contact your welding store to be sure you get the proper stainless wire for the stainless base material. An example of this is 304L base requires 308L.

Chromo : Rarely found in wire, this is one I do not recommend you buy or use. It requires certain treating and preheating processes, and for just about 99% of off-road fabrication it's not needed. For a beginner, I suggest staying away from this material as filler.

TIG welding ROD – Even more than MIG wire, the TIG rod selection is confusing. There is an abundance of misinformation regarding TIG rod usage and strengths. The main misconception is that utilizing certain rods will increase the strength of the structure. Typically, this is NOT the case.

Some rods will increase the strength of the weld, but only the weld. This is mainly true for Chromo welding. Chromo is only going to be so strong and is very susceptible to being affected negatively by the welding heat. Typically, Chromo will break right at the edges of the HAZ area of the weld. The HAZ area is the "Heat Affected Zone" of the weld and is technically the most brittle area of the work piece, unless properly heat-treated post welding. What this means is that some welding filler will not provide any more strength than others in some situations.

ER70s-6 : This welding rod can be found in all sizes and is the most economical of the welding rods. Be sure you get the -6 filler. The others are not going to be optimal for what you will be welding in off-road fabrication. This is a good rod, but not great

ER80s-D2 : The optimal rod for everything off-road. Very few utilize this rod simply because they don't even know it exists, but it's quality is better than the ER70s-6 and can be found everywhere for only a small dollar amount more. This rod works great for single and double pass welding. It can even be treated, making this a GREAT all-around off-road welding rod.

Stainless Steel : This is a little tricky. The welding rod will reflect the base material you are welding. For this you need to contact your welding store to be sure you get the proper stainless rod for the stainless base material. An example of this is 304L base requires 316 rod.

Chromo : Rarely found in rod, this is one rod that I recommend you do not even bother using or buying. It requires certain treating and pre-heating processes and for just about 99% of off-road fabrication it's not needed. For a beginner, I suggest staying away from this material as filler.

Super Missile : "Super Missile" is the buzzword in the off-road industry when it comes to welding rod, but it's completely overrated and NOTneeded for off-road fabrication. It is very expensive and can be compared to some manufacturer's 312 stainless welding rod with quite a few different components that make up the composition of the rod. This rod needs to be properly treated to gain any real benefit from it.If you are not treating the component and using this rod, then you have done nothing other than waste money. It provides NO added benefit or strength over the above-mentioned ER80S-D2 rod. If you are using it to weld a chassis up, then you are misinformed about what makes a Chromo tube structure weld strong. Sure the weld will be strong, but as mentioned above, this filler material does nothing to combat the real issue with welding Chromo material. It will look cool and you can tell your buddies you used this rod, but it's not worth the extra money. Stick to the above ER80S-D2, and everything will be more than strong enough.

My Preferred Wire, Gas, Electrode Selection for Different Applications

The wire and gas that you select will be key to yielding the best results for each specific application below. Although there are many available wires and gas combinations, these below will get you going in the right direction from the start. As for the amperage and wire speed, they will vary depending on the material thickness, welder, and wire thickness.

MIG welding plate work / chassis work Cold Rolled Steel and Chromoly Steel 0.083-0.375" Thick: This is the majority of what off-road welding will accumulate. These items included cold / hot rolled steel and Chromoly steel-plate work, such as arms, frame gusset, and mounts.

I prefer to use the Airgas Gold Mix of 75/25 with the majority (75%) argon and the remainder (25%) CO2. Even if you are not getting your gas mixes from Airgas, then you can simply ask for the Gold Mix for Steel welding. It is very common and relatively economical. The common flow rates for you to use will be 15-20 in no wind conditions and as much as 20-24 in windy conditions.

What you want to avoid is pure CO2 gas. It will be too hot for traditional off-road material thicknesses. Now for the matching wire, I prefer to use ER80S-D2 for just about all DOM, cold-rolled, hot-rolled, and Chromoly welding. The traditional ER70S-6 wire will work great as well, but I prefer to use the ER80S-D2 because it's smoother, stronger, and works great even when heat treating comes into the equation. You can get it in spools for all the above welders.

For all chassis welding and suspension-component welding where you have a max thickness of 3/16" (0.188), I prefer to utilize wire thickness of 0.030". And for thicker components such as rear-end housings with tube thicknesses of 0.25" or greater, I prefer to step up the wire thickness to 0.035" in diameter. If you are only going to get one wire for all your off-road welding needs, the 0.030 ER80S-D2 will be the best all around for weld quality and penetration.

TIG welding plate work / chassis work Cold Rolled Steel and Chromoly Steel: For those who are going to be TIG welding, both traditional steel(hot-rolled, cold-rolled, and DOM) and Chromoly, stick to ER80S-D2 welding rod. The ER70s rods will work as well, but again I just have a preference, based on my experience. It melts smoother, provides great strength, and the look is phenomenal.

To back that up, you will want to tell your gas supplier to get you "TIG welding Steel 100% Argon" gas for your TIG machine. As for the thickness on the rod, I prefer to use 1/16" for just about everything off-road. For the tungsten size I stick to the Red cap 3/32" RED tipped pieces. In the preparation section I review the methods I use to sharpen the tungsten.

As for the gas flow of the argon, I prefer to stay in the 17-22 range. Just be sure to be in a low to minimum wind location, otherwise you will blow out the shielding gas and your weld will instantly become porous. If you have slight wind, it is more than acceptable to increase the gas flow to the 23-30 range. Just remember the more you are flowing, the quicker you will use up a tank.

Sheet metal Steel MIG – Material less than 0.083" in thickness:
When welding sheet metal with the MIG machine, I prefer to step down the wire size to 0.023" ER70s, but I still keep the same GAS as the above recommendation. I do not use the ER80s-D2 for sheet metal. You can, but it will only waste money because the ER70s is going to be more than strong enough. I also prefer to pulse-weld sheet metal with MIG. Typically, the results are better. And considering that it's sheet metal, you are not going to be concerned with it being a structural weld.

Sheet Metal Steel TIG – Material less than 0.083" in Thickness:
When TIG-welding sheet metal, I again follow the same principal as MIG welding it with a smaller welding rod. Above I mentioned using 1/16" welding rod. You can use it, but chances are that you will find it difficult. I prefer to use ER80s-D2 because it melts and forms a weld puddle smoother than the ER70s series, thus making it easier to weld sheet metal. Anything you can do to make sheetmetal welding easier

I recommend it. As for the GAS, I still use the Argon like I mentioned above. Electrode – I use the Steel RED tip 3/32 tungsten and same gas flow as above.

Stainless Steel MIG: Contrary to what some say, you can MIG-weld stainless steel. You simply need to get the correct wire and gas combination to get the job done. I do this often. All the insides of GMR9 Stainless housings are MIG-welded, because it's too difficult to TIG-weld the inside. I utilize a MIG wire that is 0.030" in sizeand ER308L material to complement the ER304L material of the housing. As for the gas I utilize a mix that includes a 2% Helium to help burn the stainless filler material better.

Stainless MIG is difficult to learn. Stainless does not cooperate like the other mild steel fillers. You literally need to wrestle with it moreand pay more attention to the movement of your gun to be sure the weld puddle is controlled.Each stainless steel has specific welding applications. So be sure to check with your supplier before welding. You might need a specific wire / gas for your specific stainless base-material application.

Stainless Steel TIG: When it comes to TIG welding stainless, I prefer to use ER316L welding rod,. The size I use most commonly is 1/16" for everything above the 0.083" in material thickness, I step it down to 0.045" welding rod for sheet metal. A good example of this is the Stainless GMR9 Housing in the photo below. I still utilize the same Argon gas and flow rates as above, along with the RED tungsten. You can use others, but I have no issue using the RED tungsten. You can use the Blue tungsten, but they are expensive. I have used them in the past and only notice a slight difference with the stainless. I continue to use the 2% RED to this day.

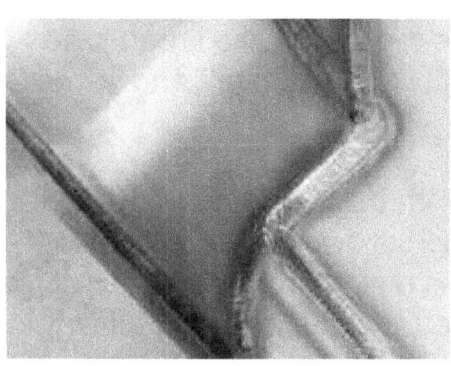

Decorative Plate work and Overlays TIG – (Silicon Bronze):
When it comes to adding some overlays and decorative plate work to a project, often you will see pros in the industry utilize a gold-colored rod: this is Silicon Bronze. It is technically not a full TIG welding rod. It's more like a brazing rod, so it's not recommended for structural elements. It burns easy and it very ductile, making it easy to use on plate work. Typically I will use 1/16" with everything that is above 0.083" in thickness and smaller 0.045" rod for thinner materials. You can also use the same Argon gas as you do for all steel TIG welding,as mentioned above. Again, this is NOT a structural welding rod, so do not use it if strength of the weld is a concern.

Metal and work piece preparation

Before you can master the techniques of MIG and TIG welding, you must first master the preparation of the material you will be welding. The four main parts of off-road fabrication welding preparation are tube / cage work, laser cut / water jet parts, raw material from suppliers, and machined material. Each style has a slightly different style of preparation but one consistent theme. The bottom line is that the material needs to be clean -- the cleaner the better. The overall goal here is to provide a mirror-like, clean surface to weld without removing the initial base material.

Tube work and Cage weld preparation – Given that just about every off-road vehicle uses tube in some way shape or form, it's safe to say that you will certainly come across tube preparation with off-road fabrication. As mentioned above, this is important, so take your time and be patient. Otherwise you will be disappointed during the welding process when you come across contaminants in the weld puddle. Once you are done notching and fitting your specific tube, the preparation for welding starts before you tack-weld them in place.

1. Remove any sharp edges from the tube so that you do not catch on the abrasives and tear up the pads.

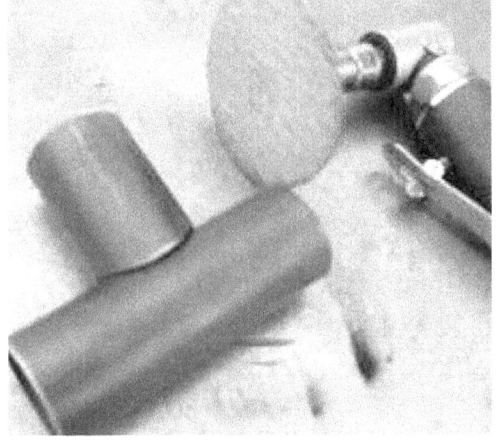

2. Take an abrasive pad like the ones pictured below and remove the "scale" of the material, the dark coating on it from the factory. This will create a clean, metallike surface. Be sure to NOT remove the base material, only the coating. This can be done with a 90-degree-angle grinder with Roloc discs, by hand with ScotchBrite, or with a belt sander and Scotch-Brite pad.

Fab Tip!
A great place to find quality abrasives at great prices is
Lehighvalleyabrasives.com

3. Once you have the material cleaned, it is a good idea to give it a quick wipe with acetone and microfiber towel, both inside and out. This is especially important for TIG welding.

When done correctly the sections should look like this -- clean weld surfaces with tube coating "scale" only on the un-welded locations.

Laser cut / Waterjet Material weld preparation – Just like the above tube preparation this style of material has its own preparation methods that I have employed over the last decade with countless amounts of laser-cut parts. From Trophy Truck link arms to daily driver prerunner rear-end fabricated housings.

There are unique issues with these types of parts because of the surfaces that are cut. The Laser cutting process leaves a small heat-affected zone on the material by the cut, along with carbon deposits on the cut surface. This creates a "scale" that will lead to weld surface contamination.

Although Water jet cutting is better, it still leaves some contamination along the cutting surface. This contamination often is in the form of surface rust (particles in the water used to cut the material at high pressure can embed itself into the surface of the material along the cut) and will also contaminate a weld puddle. The solution is simple – clean, clean, and more cleaning!

1. You need to utilize an abrasive to carefully remove the "scale" and surface rust from the material along the cutting surfaces, so that it will not contaminate the from the material along the cutting surfaces, so that it will not contaminate the degree angle grinders with Roloc discs, or a belt sander.

2. Then remove the scale from the top / bottom of the material surface along the weld surface. This can be done with the same tools as above.

3. Lastly give the material a quick wipe down with a rag and some Acetone to ensure removal of any outside contaminates.

This is the process that all the high-end shops and fabricators utilize to achieve the glorified "look" that is associated with so many suspension laser/ water-jet-cut components.

Raw Material from the Steel yard – This material type includes everything from hot rolled strip steel to the new pieces fresh off the truck. The reason they are different from the above laser cut material is because they do not have the "scale" along the cut edge or surface rust from the water used to cut the material.

Raw material from the steel yard will come in a "pickled and oiled" form with a coating on the material to keep it from rusting while at the steel yard. It can be seen in the sections above. Bottom line is that you need to remove this coating from the weld surface before welding. Removal of the coating can be done with the abrasives mentioned above, just be sure to NOT take any of the material away from the base material.

Hint –

Hot rolled steel will have more "scale" on it, while cold rolled steel will be easier to prep and will weld better!

Machined Components – As the progression of off-road has involved more and more CNC components, like billet machined kingpin beam-ends, the welding of machined parts has become more common. Just about every off-road vehicle build involve the welding of machined components. Typically, machined parts will already in the raw material format with no "scale" on the welded surface or component.

What machined parts do have that the other materials don't is machine oil. This simple liquid can be very frustrating when welding. It will contaminate the weld and cause porosity, along with very poor weld quality. The key to fight this is a bucket with some Acetone in it. Dip or soak the part in the acetone for a few minutes to remove any and all contaminates. Then thoroughly wipe the part down with a microfiber towel. Take your time with this, especially when TIG welding machined components.

Another Useful Tip! *– When cleaning bare metal, whether it's any of the above materials, you can always use some good old-fashioned elbow grease and Scotch-Brite pads! Cheap and very effective for just about everything! For a more powerful effect you can always try strapping that pad to the bottom of an air powered DA-sander!*

Preparation Bonus Section – Frame Rails!

As a little bonus to the preparation, I have decided to cover the preparation of frame rails (stock chassis parts), simply because all too often you will need to weld to a stock coated chassis. Possibly the worst of the coatings will be a stock Silverado chassis coating. Either way, all chassis need to be heavily prepared before welding or any fabrication can take place.

The coating on the frame of a truck will not simply come off with the above methods mentioned. You will need to break out the torch and quite a few rags. The key to this is to heat up the coating just enough to easily wipe it off without catching it or the truck on fire. This is very dangerous and should only be done under the proper supervision and preparation. You will need a fire extinguisher nearby, along with some rags. Melt the coating, and work a 6-8" section of the frame at a time. Be patient and do your best to not burn yourself. I have burned myself on multiple different occasions when cleaning frames, but that was due to rushing the process. *(Below is a picture of a freshly prepped Silverado Frame)*

Once you have removed the coating to bare metal and it's cool to the touch, you can then use a simple gasket remover disc along with an acetone rag to final prep the frame surface. The cleaner the frame the better your overall fabrication will be. The reason you can't use the abrasive discs first is because they will clog up instantly and thus render the discs useless. It might work a little, but it would take well over 100+ discs, costing you money and time.

Fire prevention and Safety – Even though it seems like this is a given, I still need to review it because fire is the #1 cause of unwanted damage. Whether you are welding on a table or inside the cab of a truck with some or all the interior still inside, it's always good to keep a filled fire extinguisher within arms' reach or right next to the welding table. It needs to be close so that you can grab it quickly in case of fire.

Welding fires are typically started from sparks or from getting a section of material too hot that has hidden or under coatings of flammable material. One simple trick that I used quite often was to keep an air nozzle along with 100+ psi of compressed shop air available at all times next to me. Being able to simply blow out a little flame will quickly prevent the fire from growing. The last thing you want to do is actually use the fire extinguisher. It's a last resort, because the mess will certainly ruin the work piece and add unwanted clean up time to a project.

Types of Welding Joints

In the off-road welding fabrication industry there are two different materials used with different types that pertain to each material. The two main materials are plate work (sheet metal) and tube work. As for the plate work, there are 4 main types of joints you will encounter.

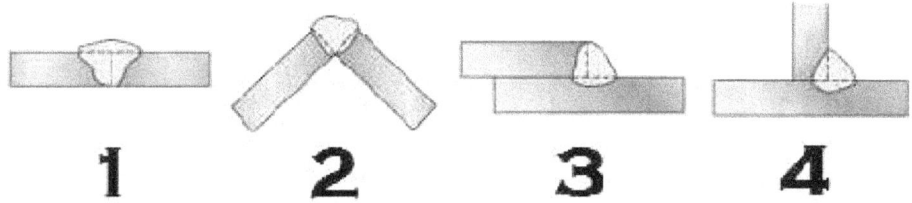

1. The first joint you see above is considered a "butt" joint, this meaning that the plates are flat or close to flat to each other and literally "butt" up to one another to create the joint. This is rarely seen in the off-road industry but can be found in some components like link arms or frame rail plating. The key to prepping this joint is that you bevel each edge and be sure to penetrate the joint fully for strength.

This joint is technically the weakest, and when seen in the off-road industry, it's commonly covered by an overlay plate.

2. This is often called a "corner" joint but can also be referred to as an "edge" joint. Either way you will see it all throughout the off-road fabrication industry, frequently on arms and suspension components.

3. When you see overlapping plate work often called "overlays," you are looking at "lap" joints. This is very common in the off-road industry and is used to provide extra strength to joints, plate work, and components with high stress areas.

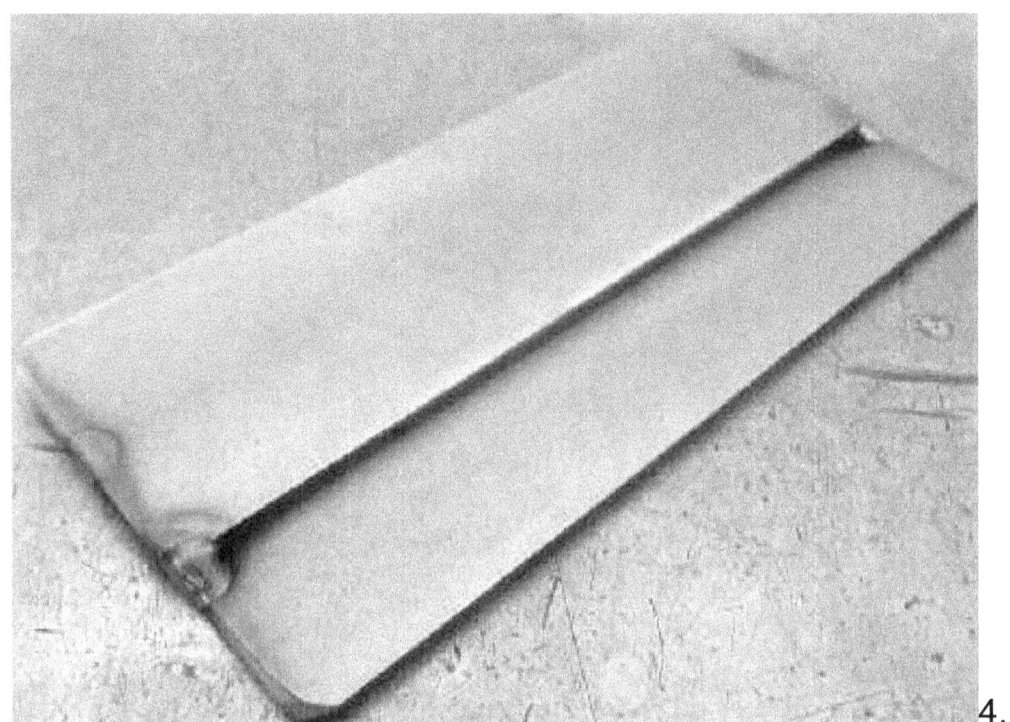

4.

The fourth type of joint is a "T" joint or a "Fillet" joint. This is often seen with laser / water jet material on arms, uprights, and many other components.

To help you understand the different joints and their place in the off-road industry, I have color-coded a few examples below that will help explain the different junctions.

RED – Corner / Edge Joint

BLUE – Lap Joint / Overlay Yellow – Butt Joint

GREEN – Fillet or "T" Joint

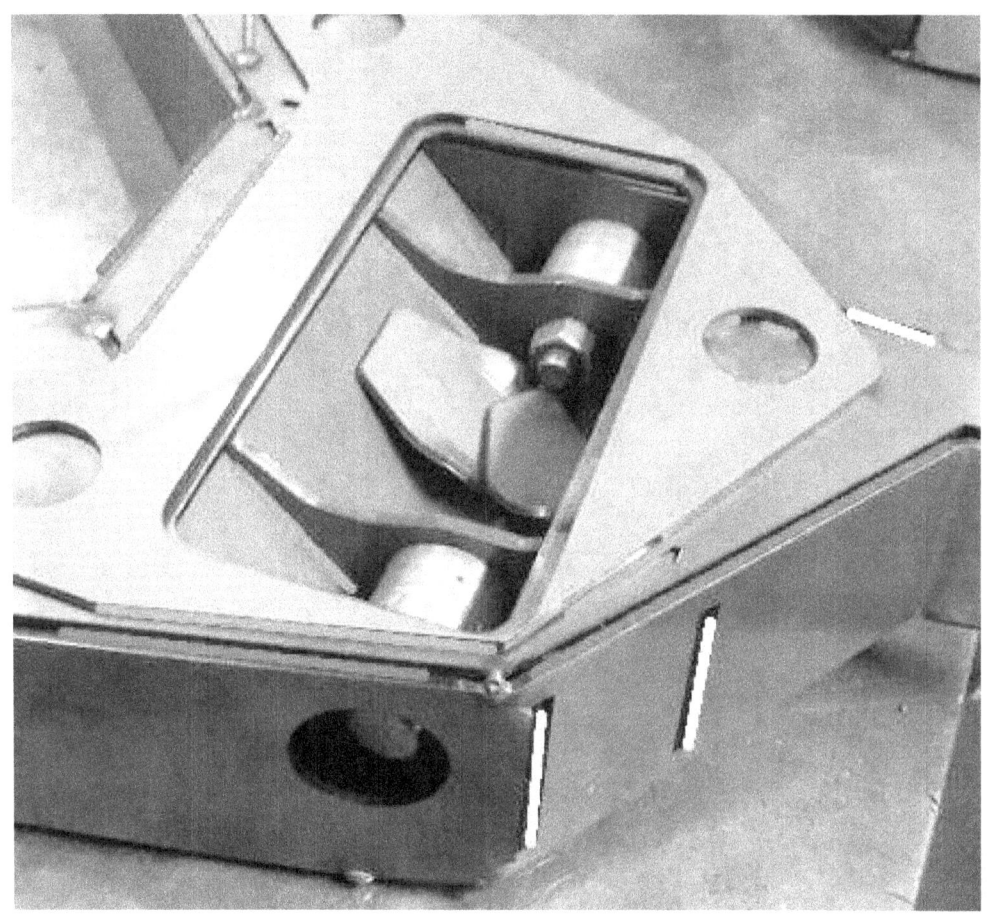

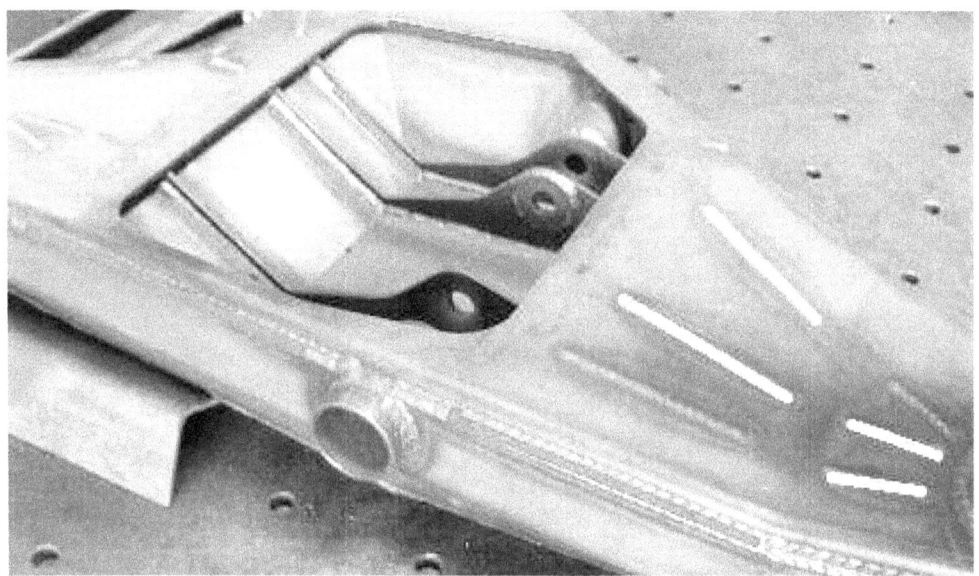

There are only two different types of Tube joints seen in the off-road industry: the Miter Joint and the Notch Joint. The Miter Joint is where tube is cut on a flat plane and either connected to another tube or plate. Below are two examples of Miter Joints. The first is with tube to tube; the second is Tube to plate.

Fabrication TIP – when joining two tubes together with a miter joint, it is strongly recommended to add a filler plate between them. This will drastically increase the strength of the joint, see below.

The above is a sequence showing the proper way to internally brace a miter joint. First you cut the miter and double check the angles. And then you make a plate (use the same material and thickness as the tube; example above is 2 1/8th" chromo tube and plate). Then you fully single pass weld the plate to one tube. After that you place the tubes together like Picture 3 above and complete the joint. Picture 4 is the complete welded joint.

The other tube joint is considered a "notch," and this is where the material is removed in the shape of the adjoining tube. This can be done various different ways, but the end result is the same -- a 360-degree fitment where one tube wraps around another. These notches can either be very long (like the below picture to the left) or simple and 90 degrees (like the below right picture).

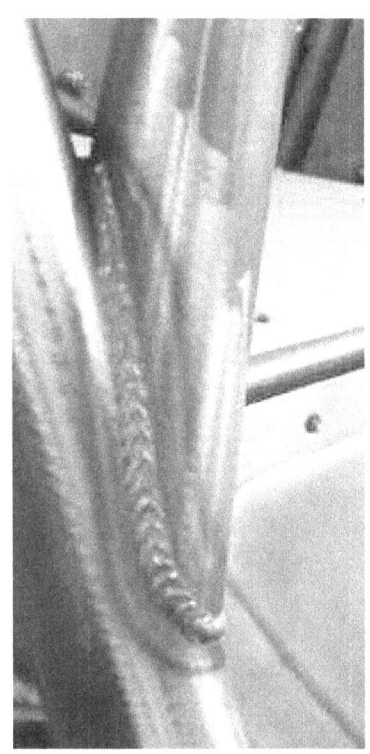

Just like the plate-work examples, tube joints also integrate multiple different joints in one tube notch.

RED – Corner / Edge Joint
BLUE – Lap Joint / Overlay
Yellow – Butt Joint
GREEN – Fillet or "T" Joint

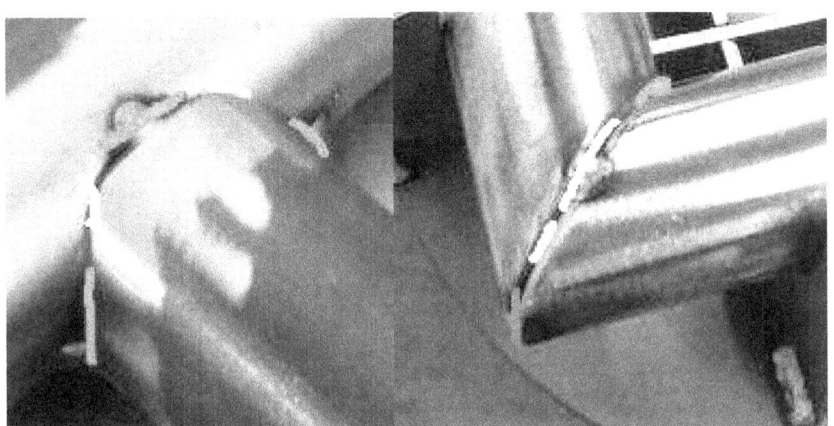

You will notice that each joint has a combination of different joints in it, thus requiring you to rotate your wrist while you weld and place the MIG gun / Torch accordingly. These will take practice, but over time you will be able to master these joints!

MIG Welding Joint setup with GUN position.
MIG welding is very straightforward. You literally set up the welder. You then locate your gun and pull the trigger.

Before you pull the trigger though, you need to know how to locate the gun properly to get a good start with your weld bead. Otherwise you could be making things harder on yourself. Below I review the recommended starting position with the joints that I mentioned above. These are only starting recommendations and will take some modification with different angles and your comfort. Each person welds slightly differently. So do not be concerned if you find it better and more comfortable to utilize slightly different angles than I mention below. Also, all the images below are designed for a right-handed welder. So if you are left-handed, reverse the pictures that reference the gun lean relative to the welder.

The Butt MIG Joint – You will start with the MIG gun leaning toward the direction of the weld path. This is considered to be a "pulling" type of weld motion, because the gun leans toward the direction of the weld, thus pulling the bead along the weld. This is the typical angle I use for the below "loop, swoop and pull" technique. The Gun should be about 15 degrees leaning toward the weld direction and about 5-10 degrees leaning in toward you. Tßhis leaning inward makes it easier to watch and control the puddle as you weld. Notice in the picture below and to the right that the gun is leaned in toward the welder. The welder is right handed and located to the right of this picture for reference.

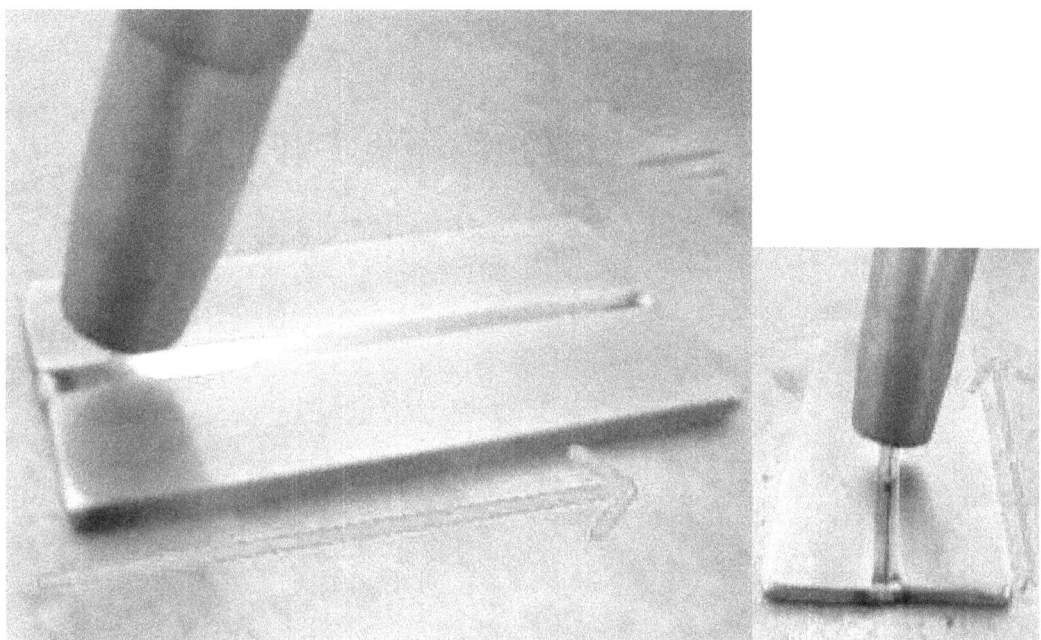

The Edge MIG Joint – This joint is very similar to the above. Lean the MIG gun in the direction of the weld but slightly in toward the welder.

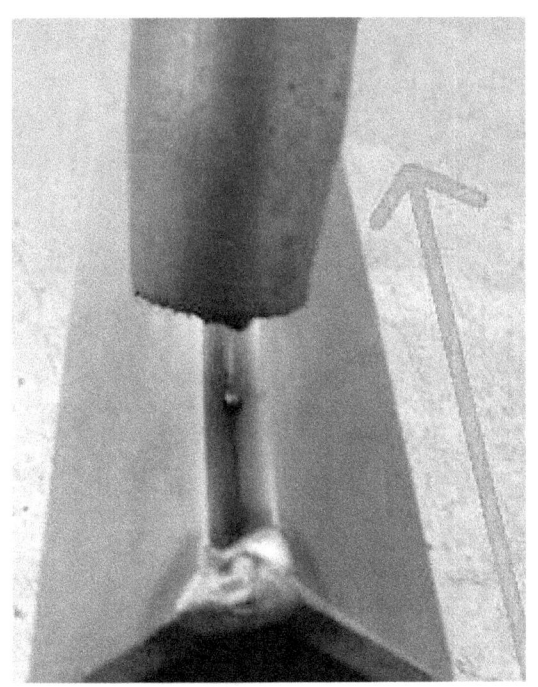

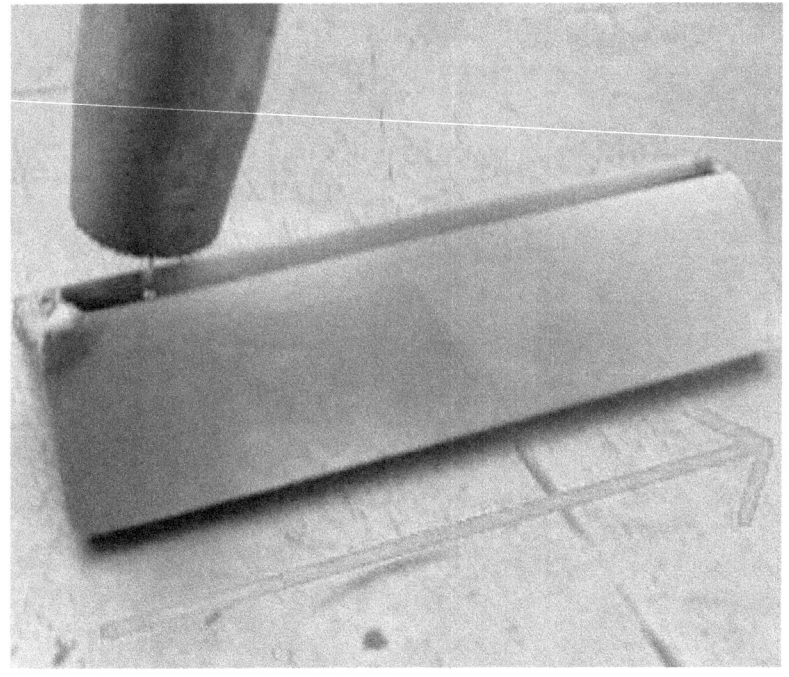

The Lap MIG Joint – You want to position the gun leaning about 15 degrees toward the direction of the weld path, while splitting the angle of the overlay. In the below right-hand photo you will notice that the gun is about 45 degrees in relation to the lap weld. This is the optimal position. You may want to experiment between 40-50 degrees to find your specific spot.

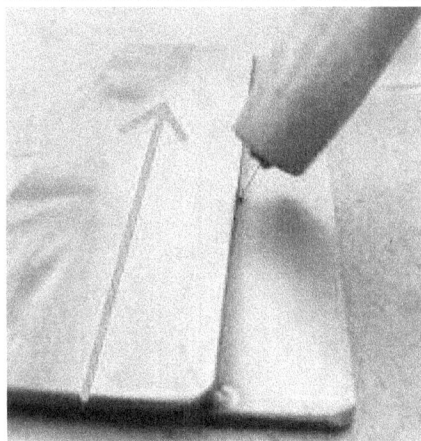

The Fillet MIG Joint – The Fillet Joint shares the same attributes as the Lap Joint in relation to the gun position. Lean slightly forward in the direction of the weld path and split the angle with about 45 degrees lean inward.

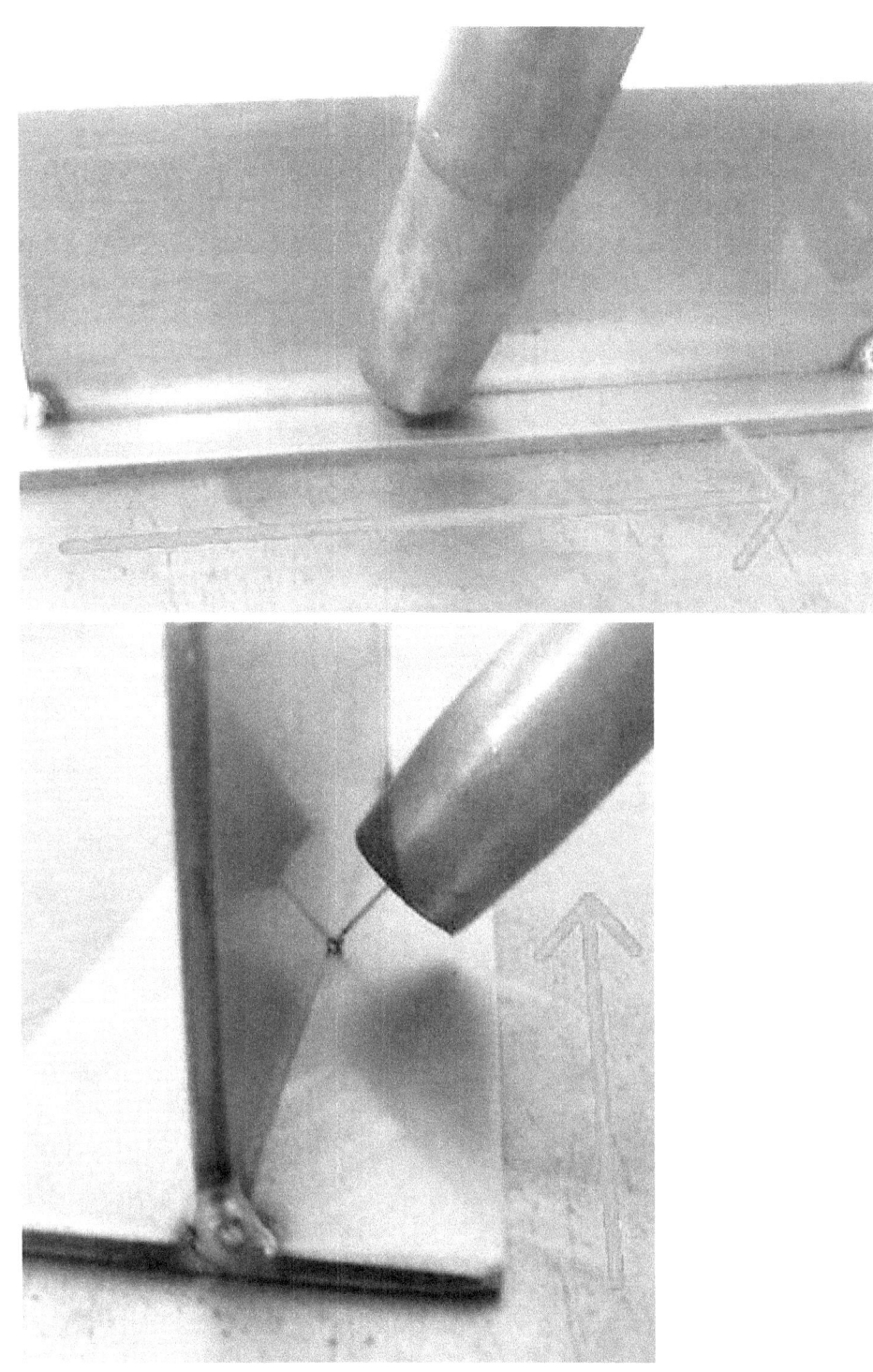

Vertical MIG welding – One rule you need to follow is "Always weld downhill. Never weld uphill." With all vertical welds and any weld that is more than 10 degrees, you will always weld downhill. Notice the path of the weld direction below along with the lean of the gun to favor the weld. This happens to be a Fillet weld, but the rule applies to all style of welds when a vertical weld direction is present.

Distance of Gun Nozzle / Tip from Weld bead – When welding any of the joints above, you will utilize the same average distance of about ½" from the welding bead. This means that while you are moving the gun, you want to keep a consistent distance from the weld puddle. If you are too close, you run the risk of contaminating the weld puddle with the nozzle / tip. When you get too close, you also damage the contact tip, causing you to replace it more often. If you are too far from the weld puddle, then the shielding gas will not properly cover the weld causing the bead to become porous (weak). If you start to see the weld crackle and bubble with little holes in it, then you are either too far or lacking proper gas flow. Check the regulator and distance from the weld puddle. When you are welding cage work and tricky sections it's OK to be farther away from the weld puddle. Be sure to turn up the gas flow slightly to compensate.

MIG welding Styles and Techniques:
The 3 Main styles of MIG welding can be seen throughout the motorsports industry but only one really stands out.

Pulse "Trigger" Welding – This style of welding is where you press and release the trigger to place single tack welding on top of each other in a row to create a bead. This method is done by holding down the trigger to place a single tack-weld down, then releasing. After the puddle on the tack weld cools, you shift the trigger over slightly and pull it again, creating another tack weld that overlaps the previous weld.

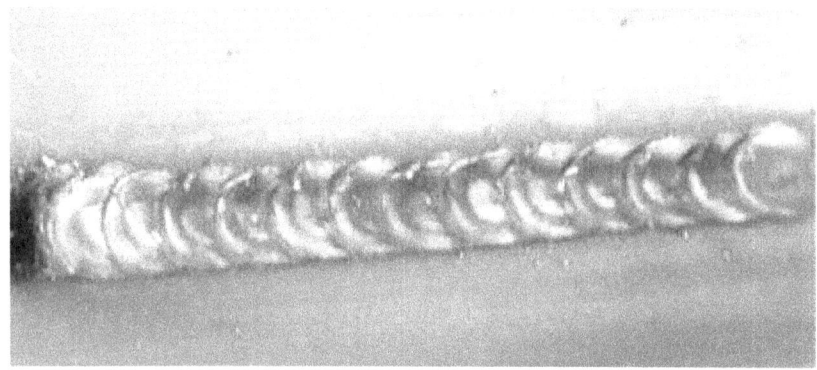

Although this is not the recommended welding technique for the majority of off-road welding, it is still useful for some lighter duty applications like sheet metal. Although some fabricators will use this for chassis / suspension welding, I strongly discourage it. In the industry, it is considered very "armature" to trigger weld cage / suspension work.

The Traditional "C" style of welding – This form of welding can be found in parts of the motorsports industry, mostly in the muscle car industry. This form of welding is very strong and has been used on everything from structural steel to top-quality racecars. Rarely seen in the off-road industry, it is typically frowned upon by the masses in off-road. The technique is easier to learn than the technique below, but being that off-road still requires some aspect of "looks" makes this the least preferred method.

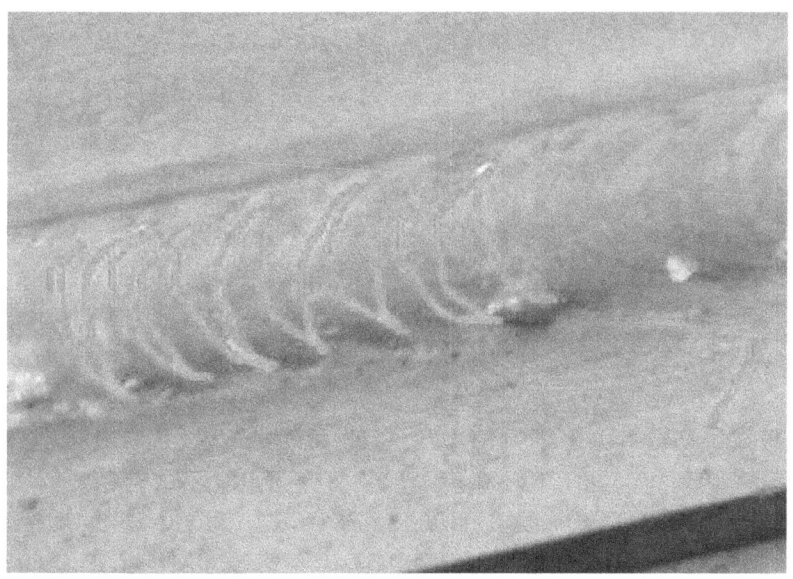

This technique is called the "c" style, because as you move the weld gun forward, the motion that you use resembles the arc of a "c". Thecurvature of the example above is not the best; it is exaggerated and a bit choppy. To reallyweld in this manor the consecutive "C's" are more refined and smaller both vertically and laterally. For the sake of example, you can clearly see what is being accomplished above.

The "Loop, Swoop and Pull" – When it comes to off-road fabrication welding, this is handsdown the best method for both strength and looks. This is also the most common method found in all of the high-end fabrication shops / builders. I strongly recommend that you focus and strive to perfect this method for your off-road welding needs.

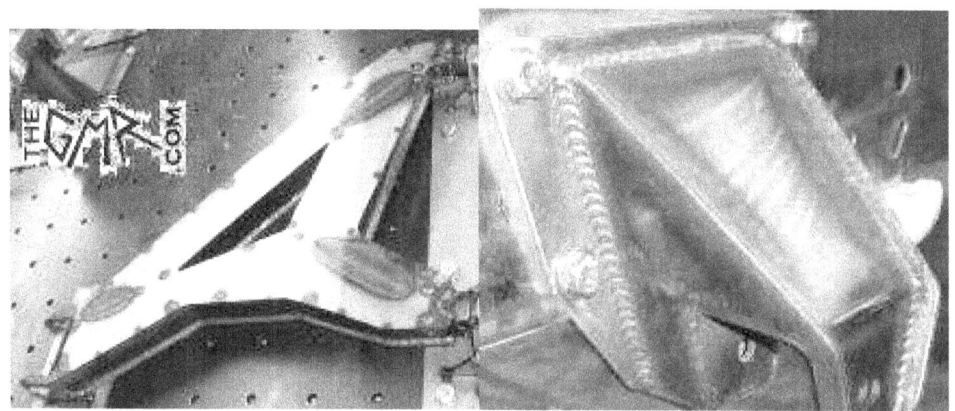

To begin with this technique, it is best to start with a simple Lap style joint that uses 1/8" or 3/16" material. The thicker material will be easier to learn on. So do not be afraid to start with the 3/16" material. Below is a great example of a lap joint on the back of a GMR9 housing that was built for a Silverado Pre-runner.

The reason this is the easiest is that you can really see the motion you are performing with the wire exiting the gun, along with having a backing (the top plate) for the weld, making this a very easy weld to control the puddle. Before you begin, be sure to set your welder to the proper material thickness settings along with prepping the gun like shown above.

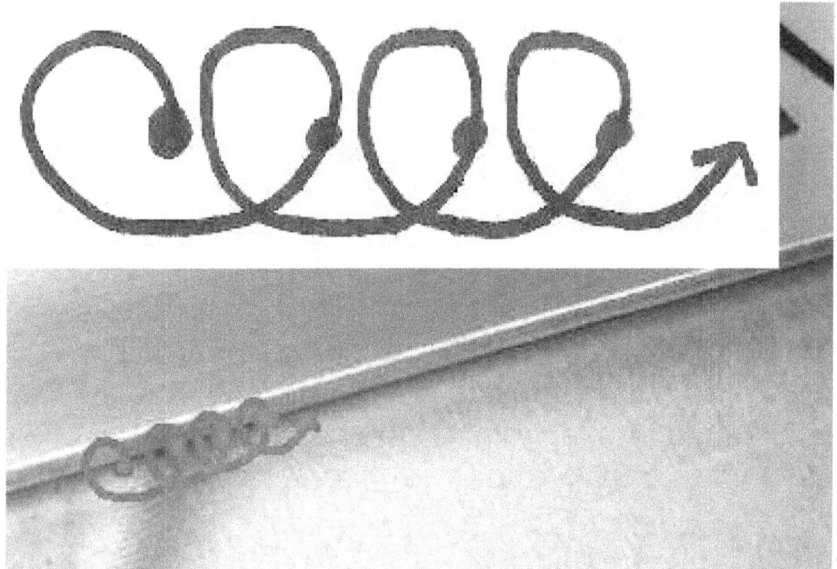

1. You start the bead by aiming the wire toward the bottom of the "top – overlay" plate and at an 45 degree angle up and 10 degrees leaning forward. 2. Pull the trigger to form the first puddle "initial start dot" for a count of "one-one thousand," then begin to move the gun.

3. Once the initial puddle is formed, begin the "swoop / loop" by moving to the top plate. Then move back in a circular motion and bring the puddle down and around, forming the infamous "dime" puddle.

4. Once the swoop is complete, you need to pull the puddle back to start another "dime". Slowly move the gun back about the same size as the initial "dime" and bring it back to the bottom of the top plate.

5. Now hold the gun again for "one-one thousand" and repeat the swoop.

6. You continue the second swoop to cover about half of the first "dime," creating a stacked look. Now bring back and repeat.

TIP – The drawings and images show overexaggeration of the actual movement of the gun. You will be making very small swoop movements. I exaggerate it to show you the technique to use. The result should end up looking like the photos below.

Get some practice under your belt with this joint before moving onto the next few joints.

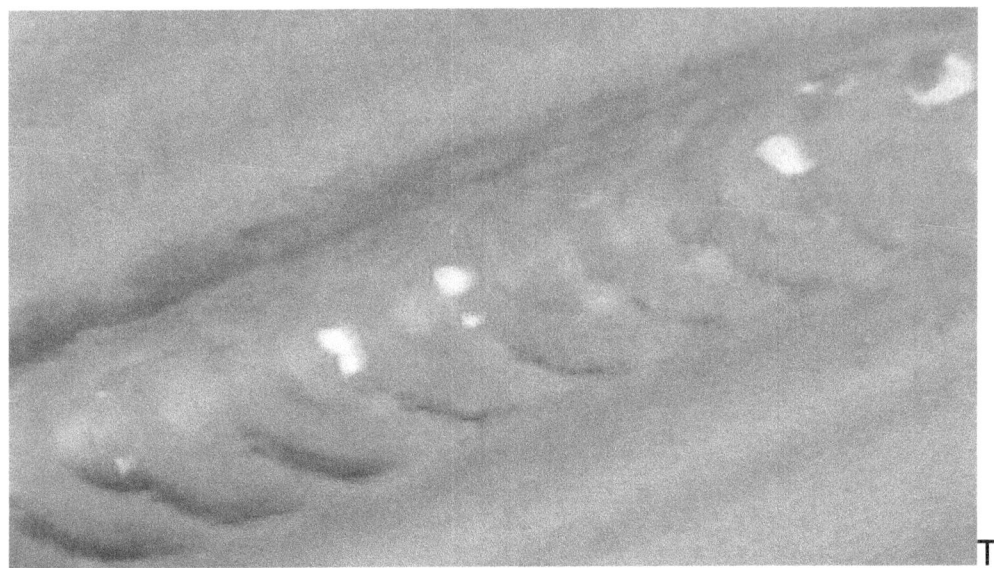

The Above Joint is the traditional Butt joint: two pieces of 1/8" plate welded together with a simple butt joint.

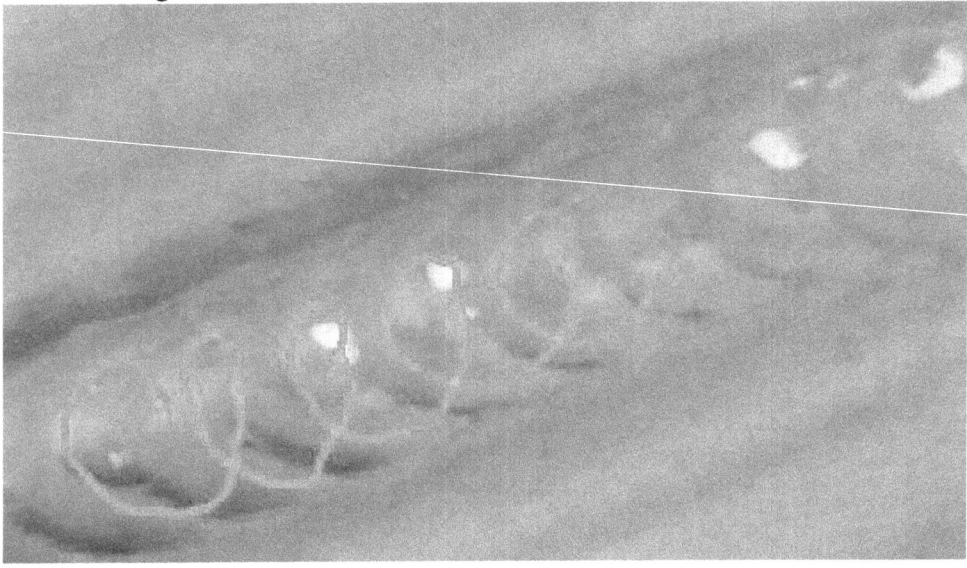

Please disregard the horribly inconsistent painting over the picture. It's not there to mimic actual movement but to display the technique. Clearly you should strive to be more fluid and consistent then the painted-on blue line.

Notice how the bead is started, then swooped and repeated to create the "stacked" look.

Below is an Edge joint. This is a little tricky and will take some getting used to. This joint is made easier by turning down the welding machine and moving slower. An example of this is the below 3/16" material that we welded with settings for 1/8" material. Slow and steady.

Notice the edges of the weld are just crowning the edges of the material. This means that there is full penetration of the material, a full 3/16" per side. The weld bead is literally a hair under 3/8" in width. Also note the stacking of the beads, a perfect 50-60% stacked on top of the previous "dime," thus creating a very strong connection between the materials.

The next joint is the Fillet or "T" Joint. This is very common in off-road and can be seen in just about everything. The key to this joint is to take your time and be sure to not undercut the material because of high heat. Check your settings and be sure you have the material thickness correct.

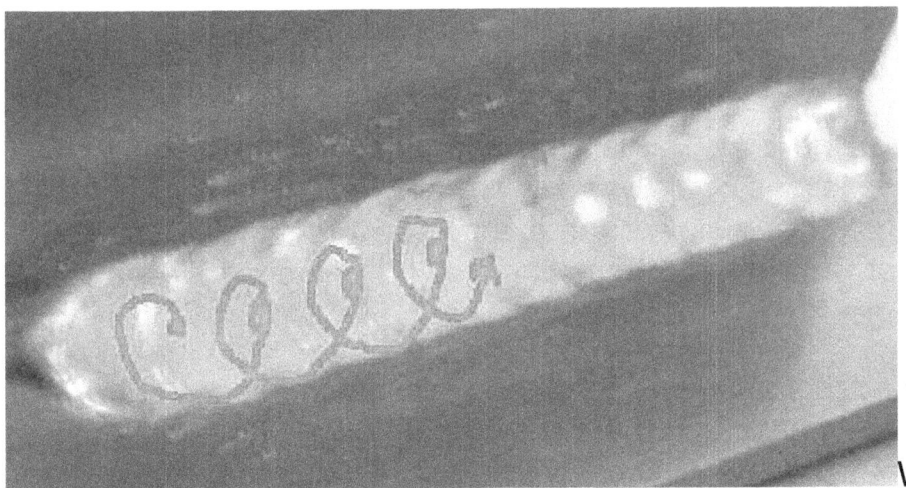With all MIG-welded joints, you want to weld from the top down, not down up. Here is a perfect example of this: a Fillet weld with the start at the top and bead continuing down.

Just like other joints, you start with making an initial puddle, then swooping accordingly with repetition. A perfect real-world example of this joint is the uncial cup weld on a race truck upright. (Pictured below)

Below is another example of Fillet Welding. This is 3/16" material to a 4" diameter ¼" thick rear-end hosing tube for a Trophy Truck. Notice the start to the top; then working the stacking the entire way down the length of the tube. This was done left-handed with welder settings for 3/16" material.

Trouble Shooting and What to Look for with MIG welding

With the above welders that I mentioned, you will find that setting up the machine will be easier to get going. The Auto-Set feature, along with the detailed instructions that come with the new welders, make the situations below less likely to come by. Either way I will review the causes and solutions to each problem.

Too Cold – The bead below is a good example of the MIG weld where the amperage is too low relative to the material thickness and wire speed. Notice how the weld bead is sitting ever so slightly on top of the material instead of melting into it (penetrating), causing a lack of proper weld penetration. The solution to this is to double-check the welder settings and turn up the amperage until optimal welding bead is accomplished.

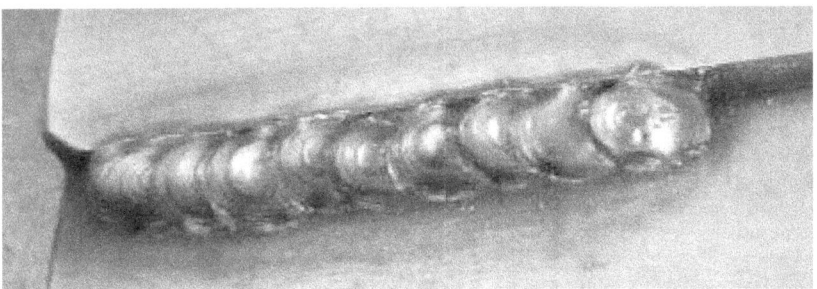

Too Hot – Below are examples of weld beads that have too much amperage relative to the material thickness and welding wire speed. Notice the "soupy" look to the puddle and the excessive heat put into the material. It's hard to see in the picture but there is even some "undercut" going on with the weld. I review what undercut is toward the very end of this book. You need to reduce the amperage until the optimal weld bead is accomplished.

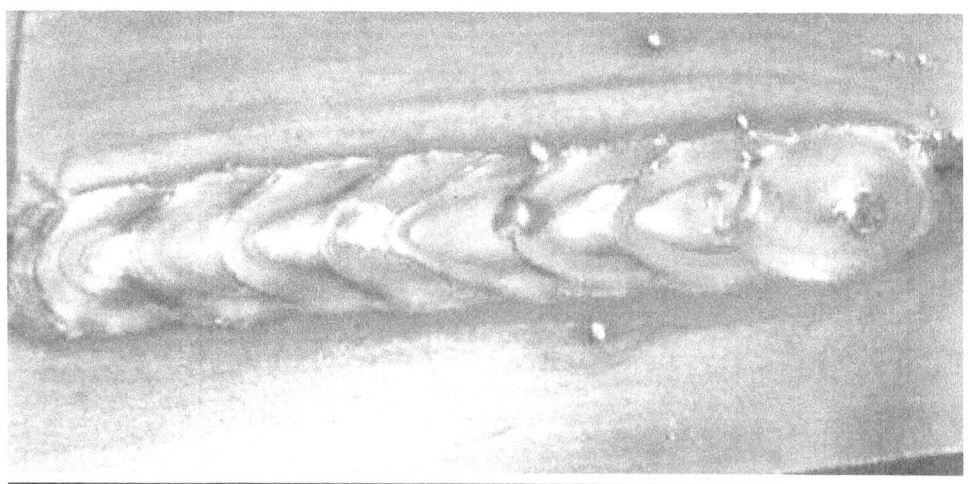

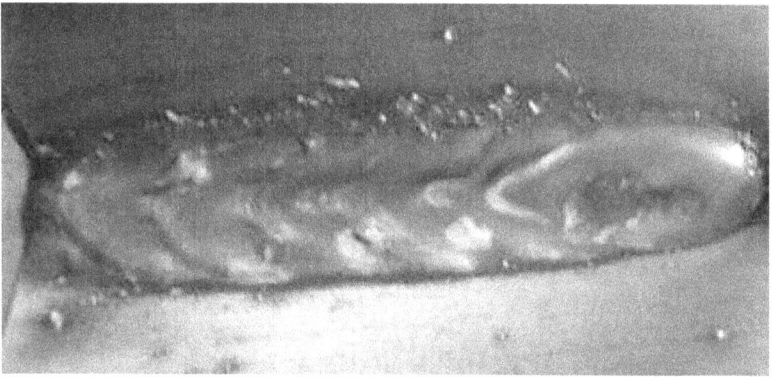

Too Slow – Below is an example of a weld that has the proper amperage and wire speed, but the movement of the welding gun forward was too slow. The bead is stacking up too much, and there is more heat going into the material than desired. The "dimes" are overlapping too much. You need to separate them more (less stacking) by speeding up your gun movement along the weld path.

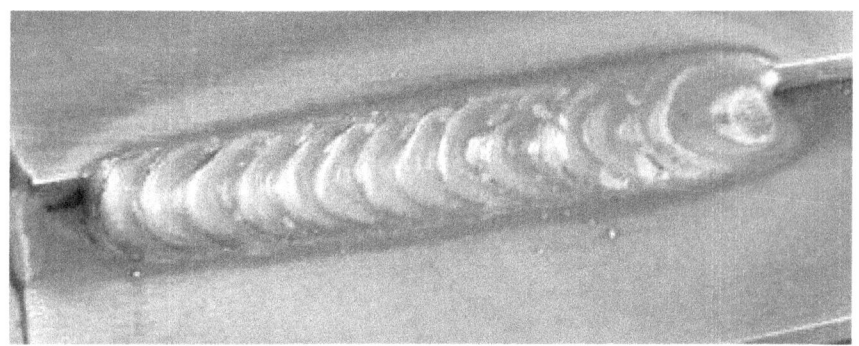

Too Fast – The opposite of the above weld. This is the same, but it's been down too fast with the gun movement along the weld path. The settings are accurate, but you need to slow down and stack the "dimes" better so that you achieve the optimal 50% range of overlapping.

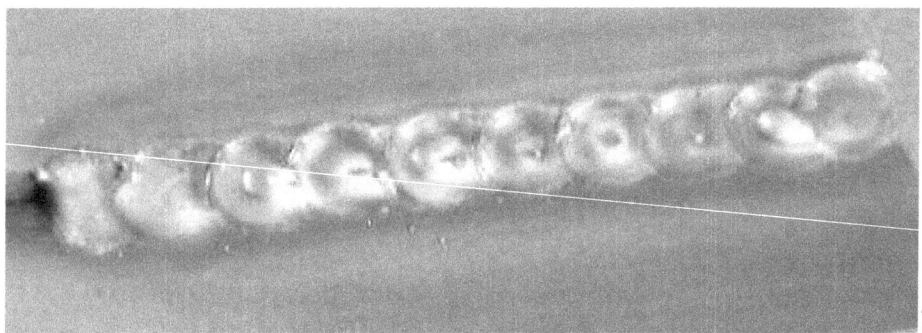

Tube welding with MIG and the "Loop Swoop and Pull" technique.

Before you start welding Tube work with MIG, you need to know how to tack weld so that you do not run into issues while welding. When you are building tube structures, it is vital that you properly tack weld each tube in place or you run the risk of a tack failure. When a tack fails, a tube can then become dislodged, causing harm and creating a large GAP that cannot usually be welded. To prevent this, I strongly suggest that you either heavy tack the tube in 2 places. Or even better, tack in 4 places, as in the photo below.

Tack welding the tube in 4 places also does one valuable thing for all beginners; it separates the tube into 4 sections, making it easier to weld up. As you progress, you can get it down to two welds, each 180 degrees of the tube, but for now I recommend you focus on 90 degrees each. Break the tube up into 4 quadrants like above.

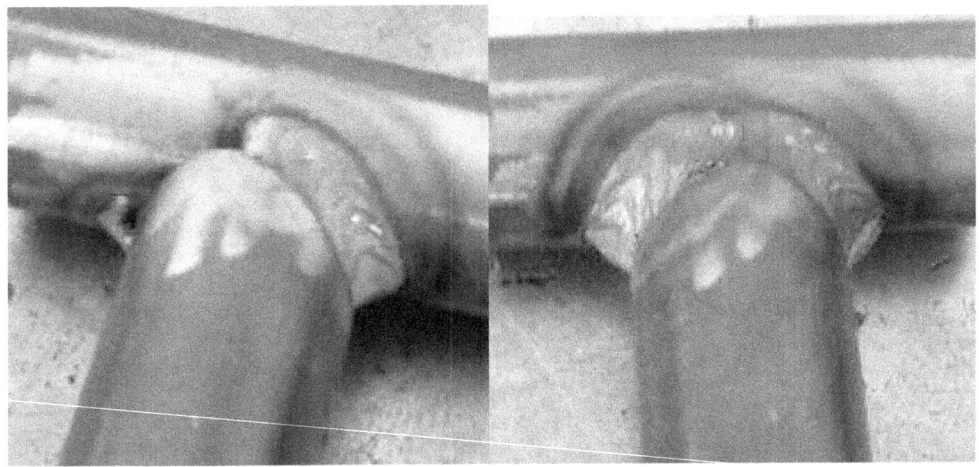

Notice that I have welded 90 degrees of the tube; then stopped and restarted on the opposing 90-degree side, completing the entire 180 degrees of weld. I repeat this on the other side of the tube to make it a full 360 degrees of weld. This method is the easiest to learn and will produce the best results until you feel comfortable to move onto 180-degree welds.

Above is another tube welding example. This time it's a tube to a flat plate, creating a true Filler joint around the tube. Similar to the above Fillet example of the Trophy Truck Housing tube, but this tube is only 1.5" in diameter to represent a cage tube. The same approach is taken with the four tack welds and breaking the weld into 4 quadrants welding only 90 degrees at a time. *Also, note the bevel on the tube to help with weld penetration.*

Below are some examples of tube-work welding in cages and chassis's. Note the directions of each bead and how they come together in some sections to make a seamless weld throughout the joint.

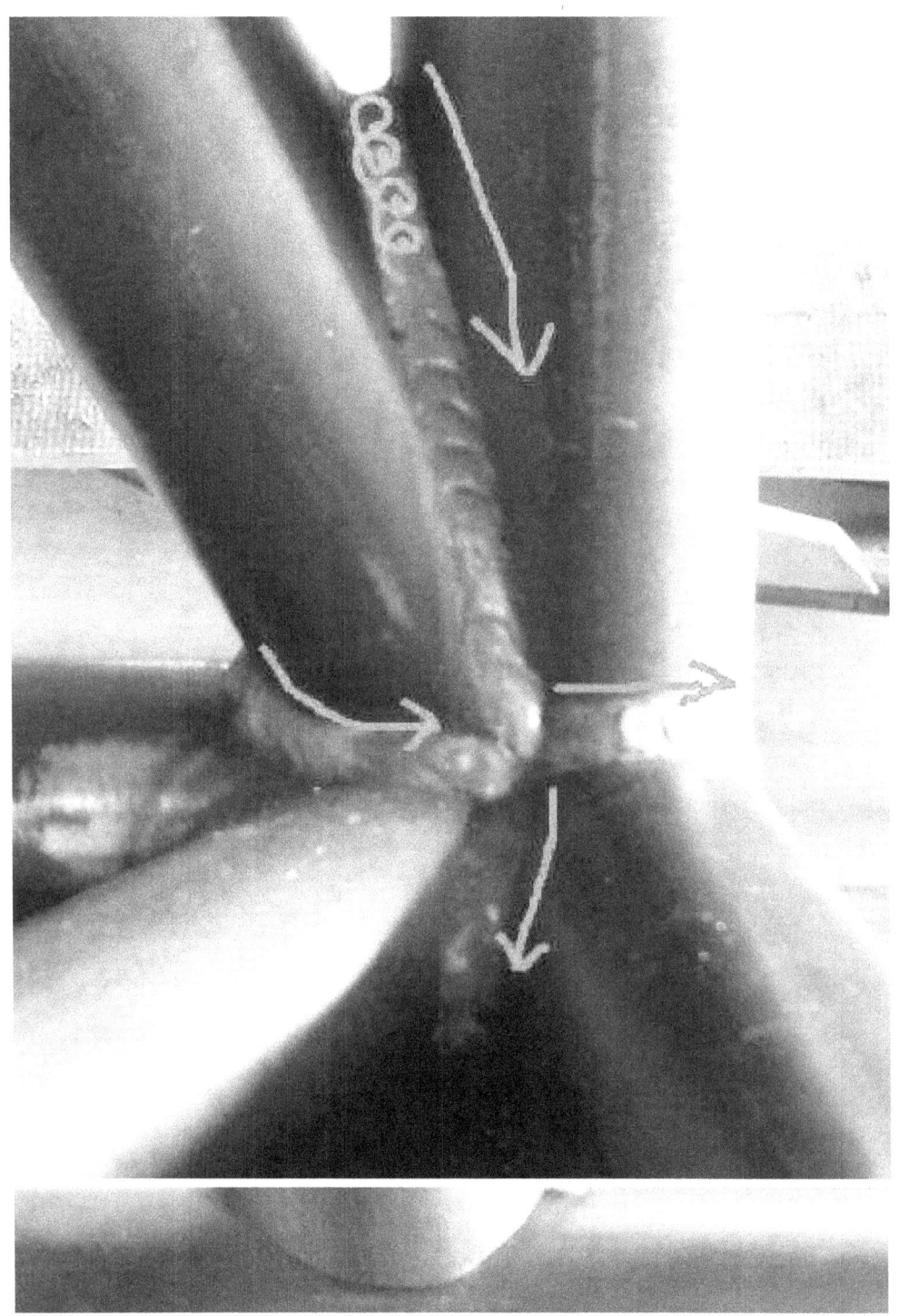

110

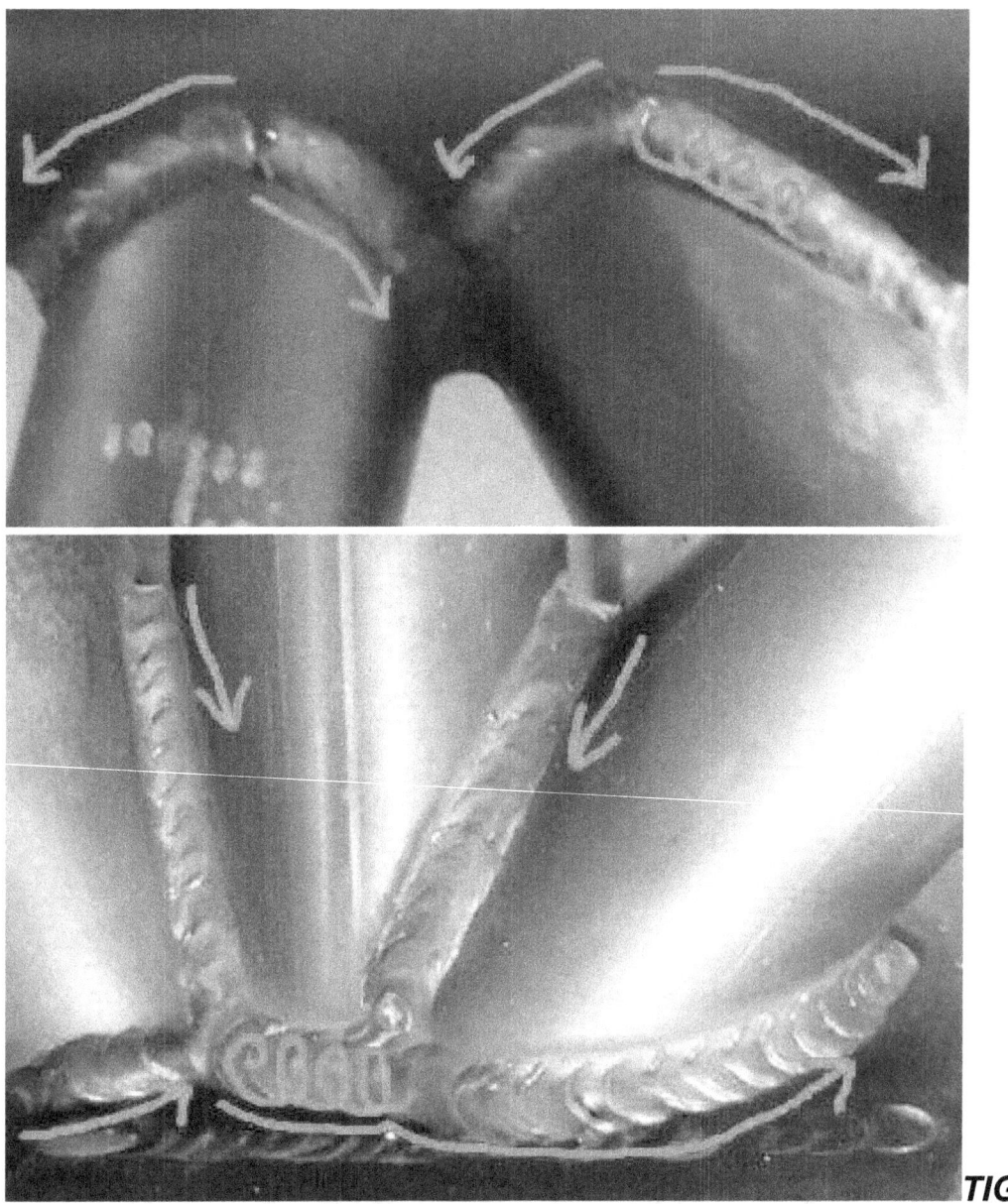

TIG Welding Joint setup with Rod / Torch position.

Before I dig into the actual techniques for TIG welding, it's important to review the proper torch and rod position to ensure proper starting technique. These are only recommendations. Once you start welding and gain experience, feel free to tweak these to suit your specific application. All the diagrams below are pictured for right-hand users. In other words, they are set up to show you the position if you were to have the torch in your right hand and the rod in your left. A general rule of thumb is that you want to all the rod to the bottom of each weld puddle. This will allow you to properly monitor the addition of filler and control its position.

The Butt Joint Position – Given that both plates are perfectly flat, you are going to split the angle, which is 180 degrees, and you will end up with 90 degrees, the standing position for the torch (seen in the image below, center). This is only when you are looking down the direction of the weld, not from the side.

In the image to the right you will notice a different angle and lean; this is the weld direction lean, or the lean of the torch in the direction of the weld. If you are welding to the left, like in the picture on the right, you will notice that the torch is leaned back, exposing the tungsten toward the direction of the weld. The general "rule of thumb" for butt welds is about 70 degrees leaning back. To complement the torch and provide optimal welding rod placement, you want to keep the welding rod 90 degrees to the torch facing the direction of the weld bead. The rod should be close to following the direction of the weld (see the middle image) but it's not critical to be perfectly straight to it, slightly off up to 15 degrees is more than acceptable.

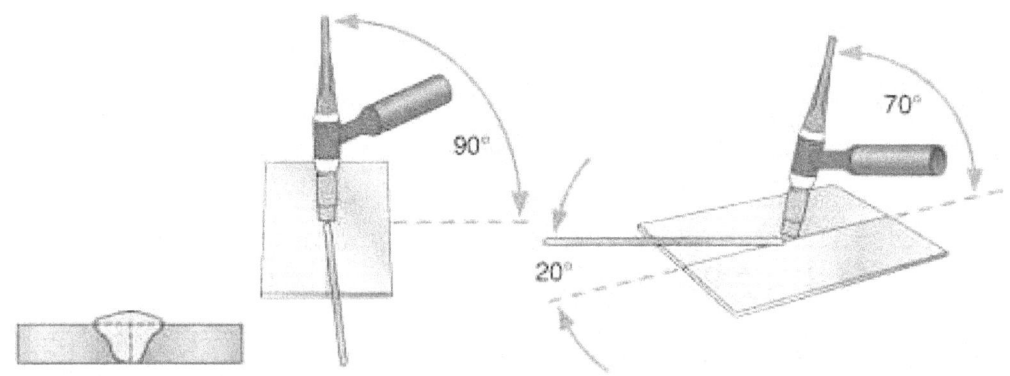

The Edge Joint Position – This is a little tricky because not all the time will you have optimal position of the work piece with this joint. A perfect example of this is when welding Aarms, and you have them in a fixture or table where you are forced to weld them standing up. One side is 90 degrees to the table and the other (top) is flat, so the joint is facing 90 degrees differently than the picture on the left below.

You will also start to see the strategy for TIG torch position -- you are looking to split the joint as best you can. If the plates were even andthe joint is facing up, you will be looking to achieve a 90-degree angle on the torch, splitting the joint. If you are slightly off by 10 or so degrees it's not going to be a problem. The same rules and guidelines for the Rod apply to this joint as the above Butt joint.

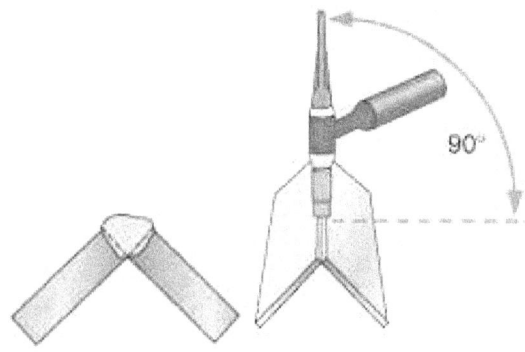

The Lap Joint Position – Breaking the pattern that you see in the above two welds, the Lap joint is different in regards to the torch position. This is because you have the full side of one plate above the top of another, thus forcing you to NOT split the joint so that you can direct more heat into the lower plate to ensure proper penetration. With the torch both 70 leaning back and 70 degrees up, you can provide the proper penetration for this weld. The welding Rod still follows the same general path as the Butt joint, but it's OK to cheat it even more than 10 degrees leaning in than depicted.

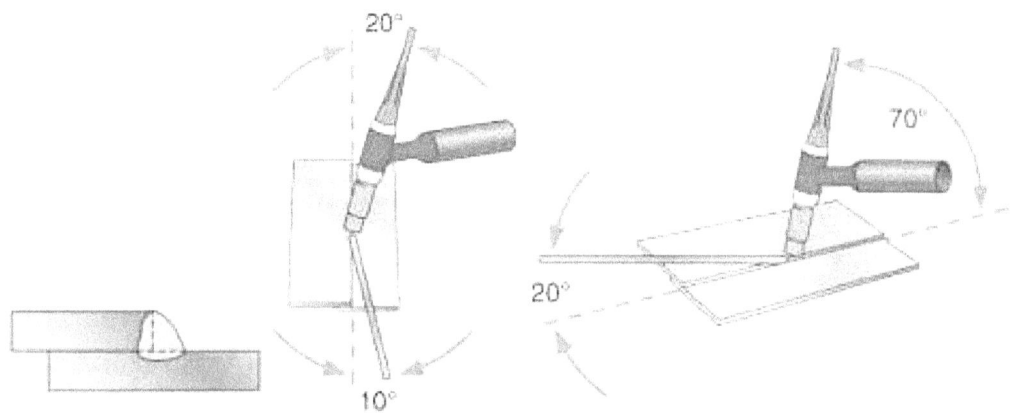

The Fillet Joint Position – With the idea that the plates are perfectly 90 degrees to each other, you are going to favor a degree split with the torch, favoring the bottom side. This is why you see a 40-degree angle and not 45. You need to make sure penetration is complete with both top and bottom plates. The other difference you will notice is that the rod will be at a farther angle away from the weld-bead direction. Below it's showing 30 degrees, and that is pretty close to what I utilize.

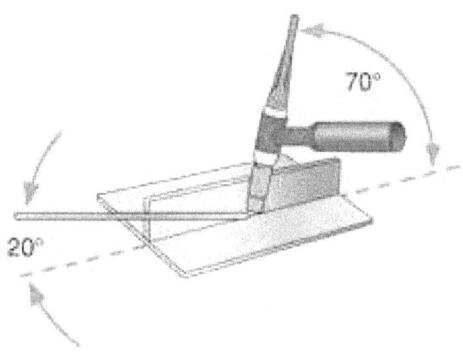

Keep in mind that all the above joint positions are simply suggestions. When you are welding in the off-road industry, rarely will you encounter perfect weld-joint position. Often you will have to adjust the angle of the torch to compensate for gravity, and watch your weld puddle accordingly. The goal is that with some tweaking of these above positions, you will be able to weld anything around or off the chassis.

Distance of TIG torch / Tungsten from welding Puddle – One of the most challenging elements to TIG welding is controlling the distance of the tungsten to the welding puddle. As mentioned above, the tungsten can be of various different lengths when it sticks out past the TIG nozzle. This is not important. As long as you have proper gas flow and coverage, then you can literally have the tungsten sticking out 1" from the nozzle (typically with a gas lens). The important thing to note is the distance of the tungsten tip to the weld puddle. The general rule of thumb is that you want it to be about 1/16" away from the puddle, but never in the puddle. Once you touch the tungsten to the weld puddle, you need to clean the tungsten because it's now contaminated. The farther away you are from the puddle, the more difficult it is to control the puddle and weld. This takes practice. Do not get frustrated. It literally takes hundreds of hours to properly master the distance of the tungsten with all welding positions, especially in chassis welding.

TIG - TACK Welding – When working with plate work or extensive tube setup like chassis work, it is critical that you properly tack-weld your work piece so that you avoid any issue while welding. Below is what a properly tack-welded overlay plate looks like.

You will notice that each tack weld is about one each from each other. This is because all too often your work piece will move and warp as you weld one section up. If the plate work is not properly tack-welded down, it will start to pull itself away from the base material causing a problem and gap with the weld. Keeping things whereyou want them is key. Be sure to tack weld your work piece properly before you start to weld them up.

TIG welding technique.

You can ask about 10 different professional TIG welders in the off-road industry about TIG welding, and you will certainly get more than 10 different answers. With that in mind, please know that what I explain below is simply MY way of doing things. There is more than one way to skin a cat so to speak, but I have had great strength results with the methods below, along with countless compliments on my TIG welding appearance. Proof…

TIG Pulse Welding – When it comes to beginner TIG welding, this form is hands-down the easiest to learn, and I recommend you start with this. The concept is simple: you are going to make several tack welds on top of each other, while adding welding rod with each puddle. This will create the stacked look you see below. One interesting fact about TIG pulse welding is that typically it will penetrate better than MIG pulse welding.

These days some TIG machines will come with a built in "pulse" feature -- I want you to turn that off immediately. It's NOT good for beginning TIG welders. It will become a crutch and only hurt you in the long run as these techniques become more difficult.

I want you to learn the hard way -- it will teach you pedal control. You need to master pedal control; once you do so then all things TIG welding will be easier.

With the "pulse" feature off, you need to set the machine to the max amperage for the selected material (see your machine's guide). Then strike an arc and bring the amps up to melt some rod and form the first puddle.

Once you have the puddle formed to size (about 3/16" to ¼" in diameter when welding 1/8" material), remove the rod and slowly lower theamperage. DO NOT let all the way off the amperage; only lower it enough to stop the melting of the rod. With the amperage at 50-60%, quickly move the torch / tungsten over to the forward edge of the formed puddle. Increase the amperage again, while adding rod to create another puddle the same size. The puddle should overlap about 50-60% of the previous one like pictured above. Repeat the "pulsing" of the amperage with your foot while adding rod each time you create a new puddle.

In the drawing above you see the red dots, which indicate the rod location for each puddle, directly in the center. The wave-like drawing below represents the amp level being increased and decreased. Notice it peaks when the red dots appear (when the rod is added to form the large blue circle puddles). Here is an example of Pulse TIG with a LAP joint and 1/8" material.

Stacking your welding beads closer together, like in the photo above, is not a problem either. This lap joint is a great example of the perfect amount of ROD and heat. Notice how the puddle perfectly lays in the joint, creating a flat, seamless appearance, even though the plate on the top left is sitting on top of the lower-right plate. There are no signs of undercut (described in the Troubleshooting Section), and each bead is very close to identical in size / spacing. I would trust this weld to a rollover.

Traditional Single Pass – The next form of TIG welding for you to try out and master is the single pass, or what I call the traditional single pass. Not everyone calls it that. Some use different names, but the technique is the same. Many of the industry's top professionals still use this technique to this day. The new Terrible Herbst Motorsports Trucks were almost entirely welded with this technique, including the suspension.

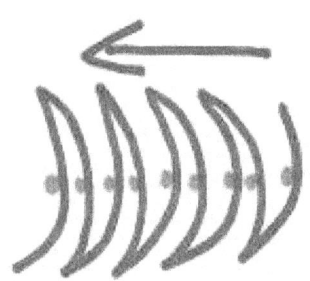

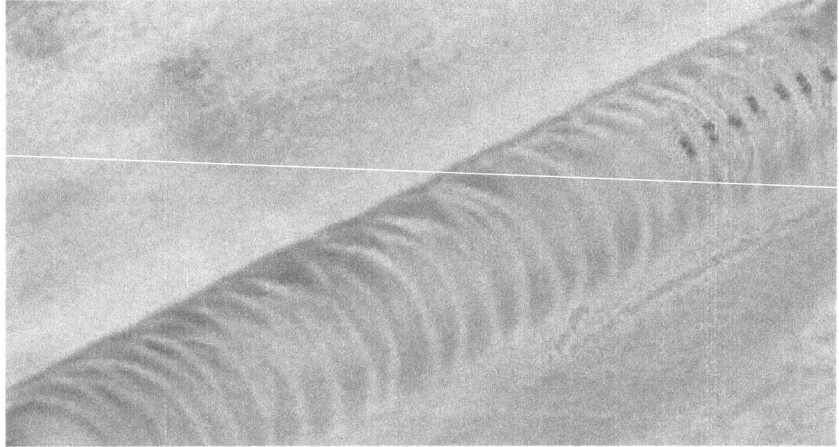

This method is simple and straightforward; it is also one of the easier techniques to learn. The key to this is consistent heat. Unlike the above pulsing technique, you will be working toward keeping the amperage consistent while welding.

The picture above is of a lap weld with two 1/8" pieces of material. I prefer to be in the 115 am range throughout the weld bead from start to finish. Different materials and positions will yield different amperages. The key is to go slow and take your time, but do not overheat the material.

As you start you form a puddle toward the base of the top material, then while adding rod work your way slightly forward and up. Once you move up and form the puddle at the top edge of the above material, you then move the puddle forward and downward, creating a "C" like shape. Notice the drawing above with the back-and-forth motion while moving forward. The red indicated where the rod should be added; when you are in the middle of the "C," you simply shove in slightly more rod than at the ends. While you move forward, you can keep the rod in the puddle the entire time and control the puddle size with the addition of rod in the middle of each "C" pass.

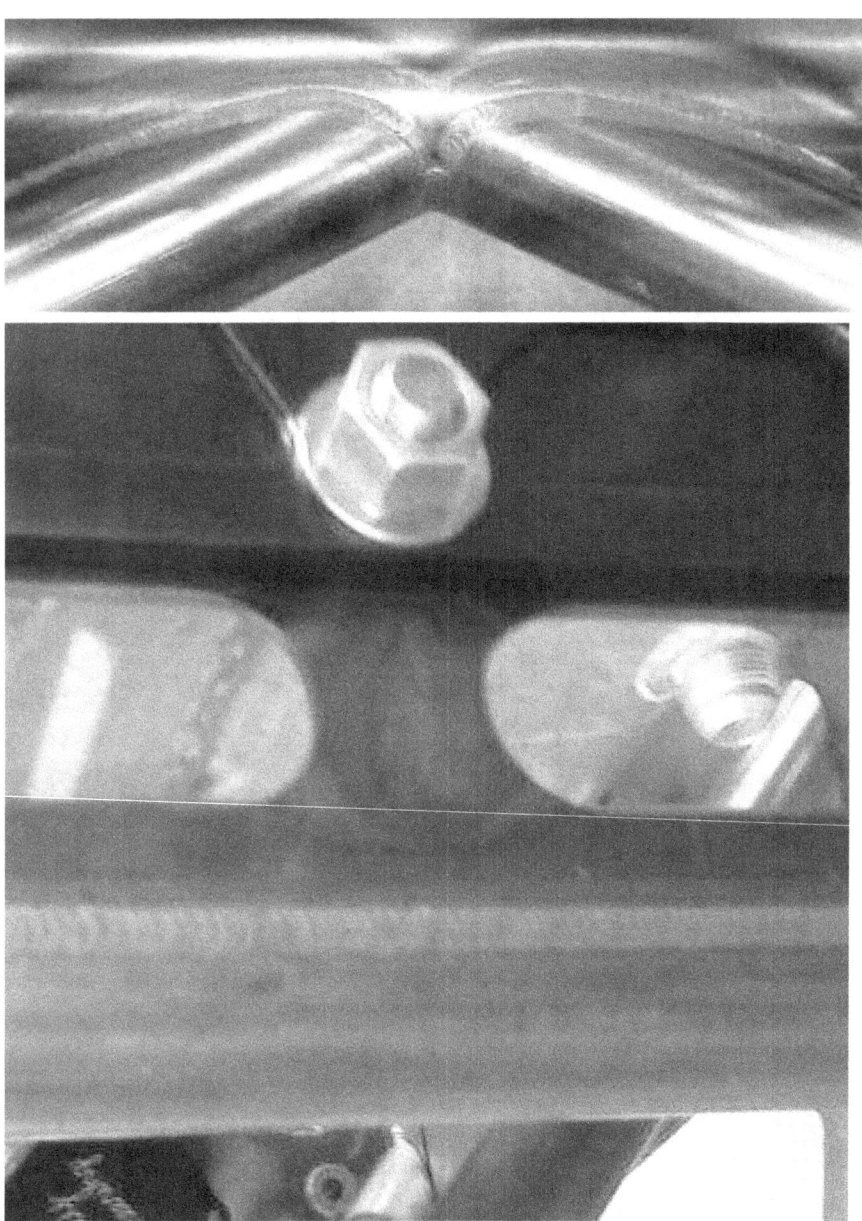

Above are perfect examples of traditional single-pass welds that are both very strong and aesthetically pleasing. Notice the symmetry of the beads and the sections where you can see the rod being added while the puddle is moving forward.

The Double Pass "Weave" – One of the most famous and notorious TIG welding techniques is the Double Pass weld. This is very complicated to do correctly and will take some practice. The top builders and trophy truck teams typically will utilize this technique for everything from chassis to suspension construction.

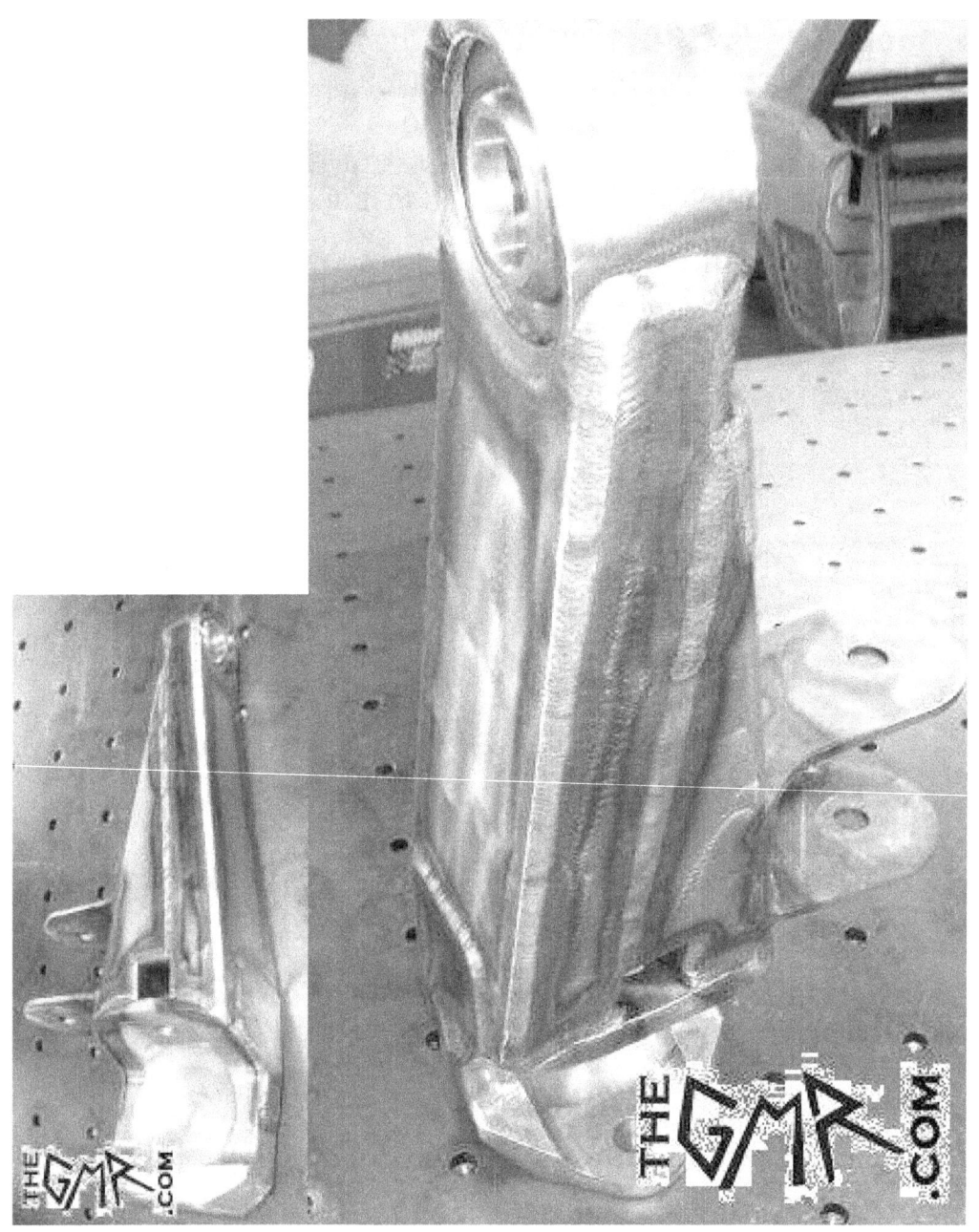

The main downside to the double-pass weld is the difficulty and time involved. If you were to double-pass a chassis (instead of doing a single pass or even MIG weld), you will be tripling the time it takes to weld the structure, maybe even more.

The first step to a proper double pass is a good foundation with a single pass weld. Notice the pictures above: the one to the left is a miter joint with a single pass and the accompanying tube tack welded in place. The middle photo is the three tubes appropriately single pass welded fully. Once cool, you can then run your double pass weld over the single pass weld. You are literally double passing the weld, hence the name.

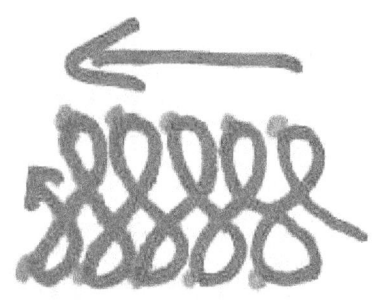

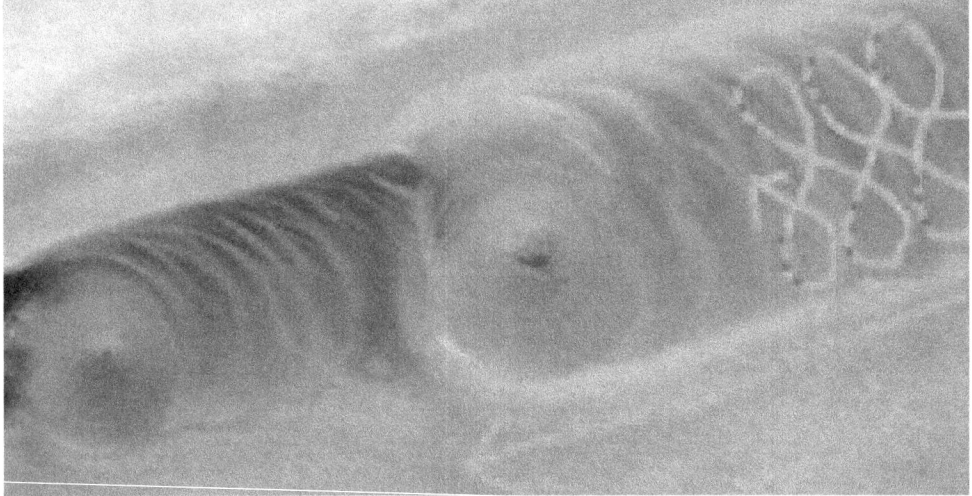

The technique for the second pass is basically a backwards Figure 8 that moves forward. Note in the drawing above the direction of the arrow. In the picture you can see the second pass going over the first with the different pattern. The reason for this pattern is that it literally blends the two materials together, creating a very strong bond. While moving forward with the Figure 8 pattern, you want to strive to keep the amperage consistent and be sure to NOT overheat the material.

One sure way to be sure you do not overheat the material is to start to weld with the amps set to below the material thickness by about 10%. So if you are welding 1/8" material (roughly .120), you want to be at about 108 amps max. This is a good starting point, and you can adjust from there, given the joint and weld complexity. The red dots represent the addition of rod. At the upward outer swoop of each

Figure 8 movement, you want to add rod. This is where you get the "dual stacking dimes" look that the double pass is known for. When you tighten up the Figure 8 pattern, you will start to see more of a "weave" going on, hence why this is also referred to as the "weave" weld.

The pattern and addition of rod should yield a result that expands the overall width of the second weld bead so that it completely covers the first. The tricky part of this weld is preventing undercut with the second pass. You will have to really watch your weld puddle and be sure you are not digging into the base material.

The picture above and on the left shows a dual pass weld joint in process. The top two welds and tube are completely double pass welded, while the bottom is still single passed, awaiting the second pass. Note the direction of the welds and how it runs over the previous welds from tube to tube, creating one continuous bead with three tubes. The picture to the right is a great example of a professional double pass weld on the critical junction of a trophy truck chassis.

It's all in the wrist . The "weave" technique will require you to practice the movement of the torch with your wrist. As you move the torch through the pattern, you will also angle it slightly inward (away from the edges) with each side of the Figure 8. This slight 10 degree or so angle will drastically help with decreasing the chances of undercut.

Walking the Cup – Another very popular form of welding in the Off-Road industry is called "walking the cup," and some use it for just about everything. I am not one of those. It has its place, and that place is not on structural components. Items like chassis and critical suspension welds should not be done with this technique. When it comes to overlays and decorative items, this technique will provide the look you are going for.

The concept is very simple. It's basically identical to the "weave" weld, but you do not add any welding rod while making your second pass. The pattern you utilize is the same; see the images below for reference.

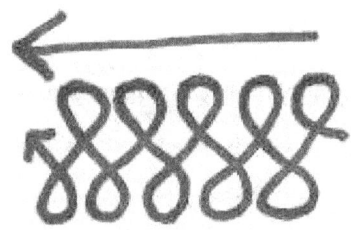

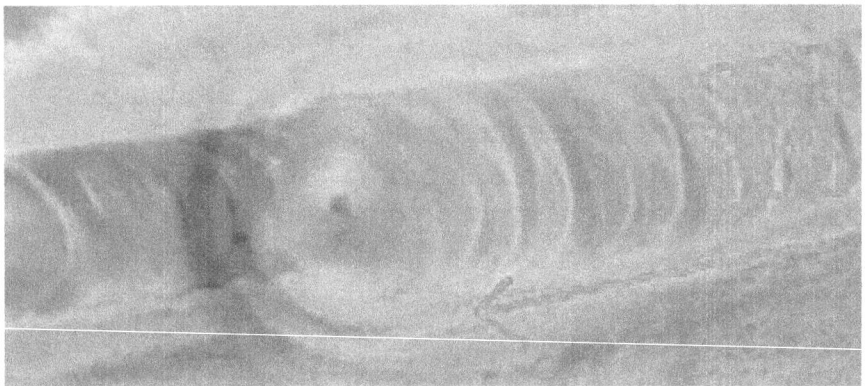

Above you can see the initial single pass weld being covered by the second "cup walking" pass. One distinct difference you will notice between this and the above "weave" is the lack of size with the second pass. The lack ofadditional filler material will yield a smaller bead (usually, but not all the time). Below is a perfect example of when this technique is more than acceptable to use. This is a logo plate on theback of a GMR9 housing that is under a 90 Firebird Pro-Tour Car.

 **UNDERCUT with Welding**

One of the worst outcomes of an improper weld is the dreaded undercut; it leads to premature cracking and failure. The pictures below show examples of undercut. Notice the edge where the arrows are pointing.

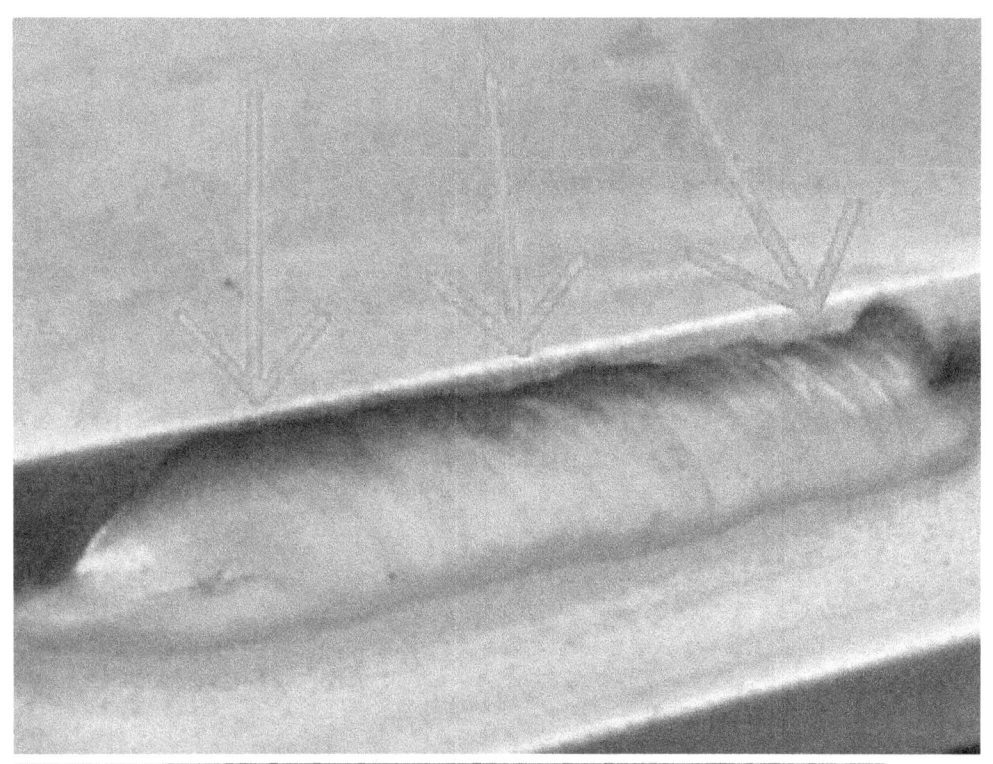

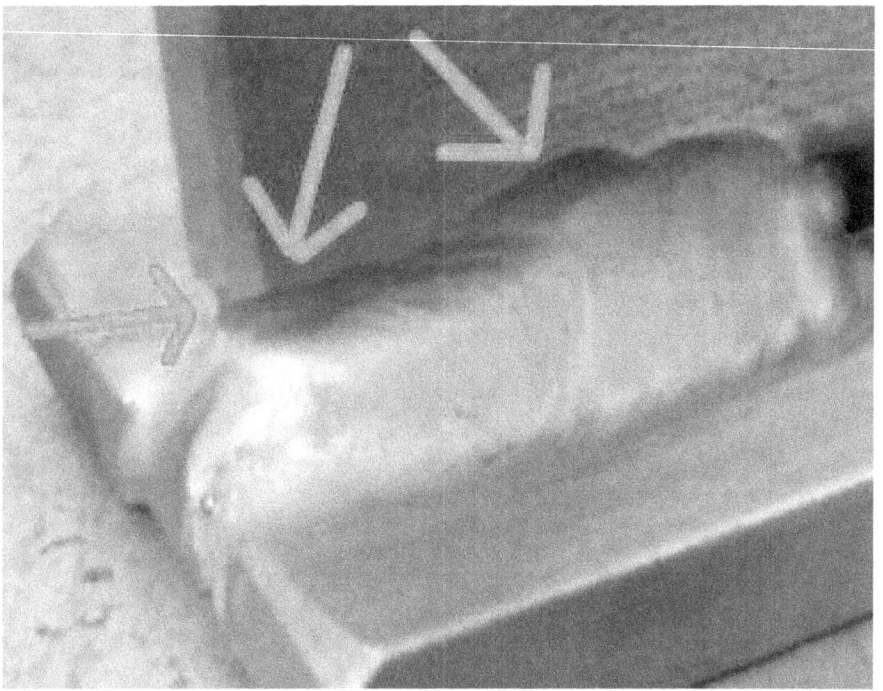

A proper weld will sit slightly above the base material, not below it. These pictures show the weld literally sitting below the base material, making the weld smaller than the base material and creating a high-stress fracture area. Undercut happens when the base material is melted into the weld bead without properly wettingup back in place with filler to prevent the edge. This can be caused by a few different things:not enough filler material; too hot, too fast, or improper weld gun / torch angle. If you see undercut happen, you need to stop the weld immediately and readjust. Then move back and fix the undercut section. Double-check all of your settings and review the gun / torch position and continue. To fix these welds, you will need to pass over the root weld with more filler and be sure to fill the void created by the undercut weld.

If you are looking for more information on build a Chassis and details around how to build a Trophy Truck Chassis you should check out my other book on the Monster Garage TT rebuild, also available on Amazon.com

Below is a little "sneak peek" of a section from that book...

After I completed the rear end, I moved onto the rear lower link arms. I chose a 60" overall length for the rear trailing arms, along with very specific design elements. They are fully constructed from 4130 Chromo. Actually, everything on this new chassis was constructed from 4130 Chromo. So if you see platework / tubework, you can rest assured that it's 4130.

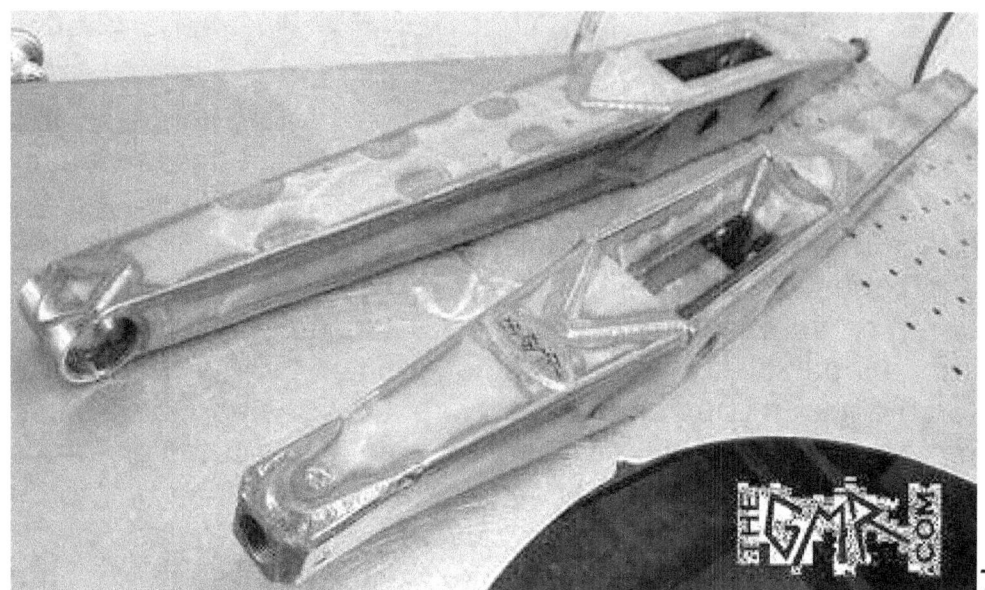

There are many different rear lower-link arm styles and designs on the market. Many of them work great, so there really is NO right or wrong way to build rear link arms. However, I can tell you how and why I built these the way I did.

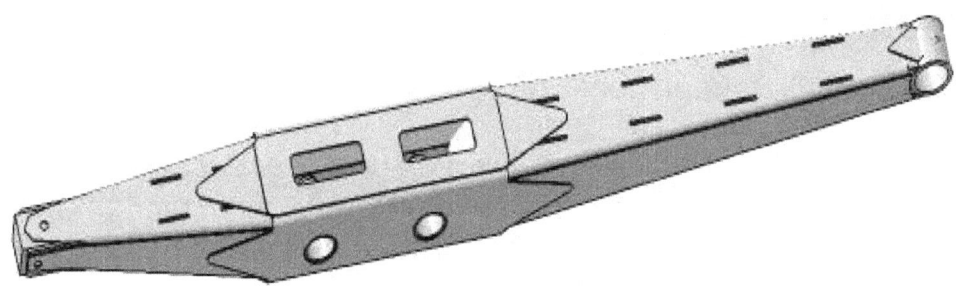

The shock mounts were selected to be in a very specific location for a reason. I wanted to provide a great internal shock piston-speed-to-wheel travel relationship. The rear shocks that we are using have an 18" stroke so I had to factor that in with the wheel travel I wanted. that we are using have an 18" stroke so I had to factor that in with the *wheel travel I wanted. 23" of travel from the front. So I decided to set the rear up to complement that, NOT simply to get the most travel we can. The suspension needs to be a compromise. And with rear trailing arms, you have to keep in mind the shock stroke length, desired wheel travel, and internal shock piston speed.*

I personally prefer to have a lower piston speed to wheel travel, which means that the by-pass shock will be mounted closer to the rear end then some builders prefer. Below is a good shot of the final shock mounts and location.
With the front at about 22-23" of travel, I was only shooting for 29" of travel in the rear.

Here is my reasoning for that. The general rule of thumb is that you want more droop and

Here is my reasoning for that. The general rule of thumb is that you want more droop and 5" more bump travel. This will give the truck good balance when it comes to the wheel-travel relationship. The coil over was then placed in a location to clear the large by-pass shock through travel, and the ratio ended up being less than 50% up the arm, giving a great relationship on the spring rates and coil-over shock usage. The size ended up working perfect with a 16" stroke coil-over. In general you want to avoid placing the coil over past 50% up the arm toward the chassis. This is one big leverage arm. And when it gets closer to the chassis, you will find that spring rates are not favorable and tuning the truck can become difficult. It's all about setting up the shocks to work as effectively as possible.

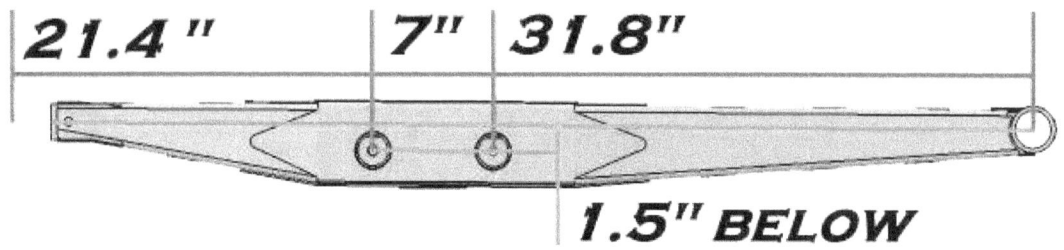

21.4"   **7"** | **31.8"**

1.5" BELOW

The last element to mention about the rear trailing arms is the shock placement relative to the centerline of the arm. Even though I'm using the wobble stoppers on the front of the link arms, it's still good to do anything you can to prevent the arms from twisting over. One very simple way to do this is to mount the shock below the centerline of the arm. I opted for about 1.5" below the centerline of the link arm, thus using the weight of the vehicle to literally force the link arm straight during wheel travel movement. If the shocks were mounted on center or above center, then it will have a tendency of forcing the arm to roll over in the truck, causing problems.

With the wheel travel in a good relationship to the front and our shock ratios placed, I was able to figure out the piston speed ratio, which was very good. It does slightly change as the shock moves because the angle of the shock changes through the travel. But keeping the piston speed as close to the wheel travel as possible will yield better tuning results and shock longevity. If the Shock stroke were to match the wheel travel, then it will be a perfect 1:1 ratio. If you have a 12" shock and 24" of travel, then you have a perfect 1:2 ratio. The higher the ratio, the harder the shock has to work to control the suspension. So by lowering the ratio, you effectively make the same shock work more efficiently and perform better. There is quite a bit more complex engineering behind that theory but those are the basics as to why I chose the locations for the shocks where they are.

Some link arms are built with a main tube construction. These are more traditional link arms, where a section of tube is used to construct the link arm, creating the base for the arm and plate work. Below is an example of a traditional tube lower link arm.

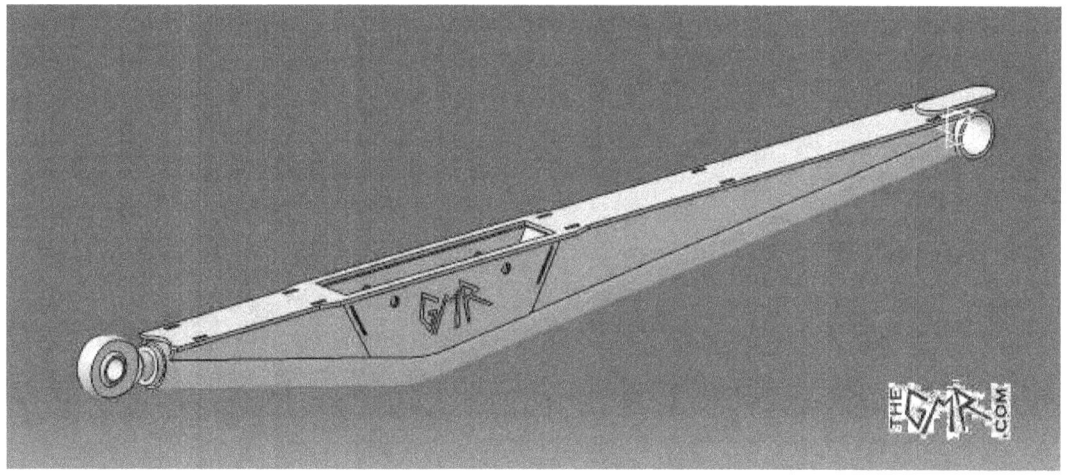

I opted to use a more non-traditional method; the link arms I designed did not utilize any tubes in the construction. Instead I used *all CNC-cut plate work that was cut and CNCformed to the CAD design I drew up.*

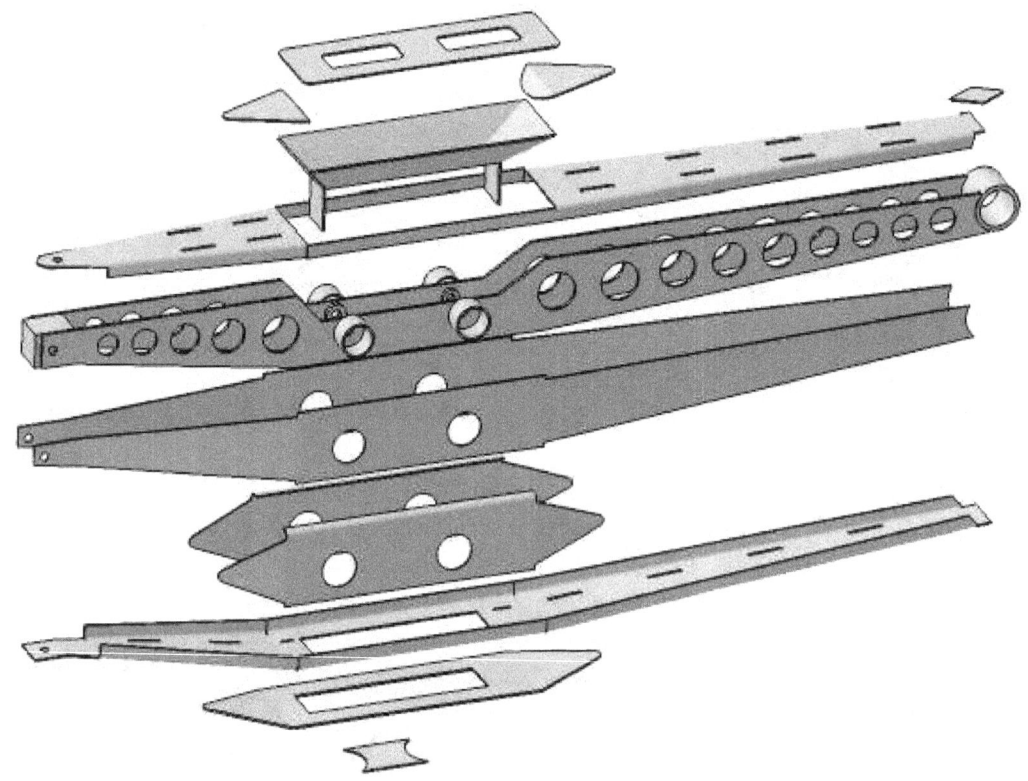

The design that I utilized features two long main ribs for the internals that were welded to each end, spanned the length of the arm, and also had the holes for the shock mounts. Then I plated in the arm around those, creating a complete box section for the arm. In the area of most stress, where the shocks are mounted, notice how I increased the size of the arm to provide the strength I required to support the forces this truck will see.

Just like the GMR9 rear end, I welded up the link arms in a complete fixture that rotated so I can weld everything accordingly. The arms were double-[pass welded with ER80s-D2 TIG rod, a personal favorite of mine. If you pick up my first book in this series, then you will notice that I prefer and recommend the ER80s-D2 welding rod. The fixture I utilized fully supported the ends of the link arms along with the shock mounts, making the construction of the arms very fast. It literally was 90% welding time to build the arms. Each Link arm took about 10 hours to weld up.

There is some debate on what is used to support the rear of the link arms and attach them to the rear end. Some prefer the use of a uni-ball while others like myself prefer to use a 1.25" rod end. Both sides have great points. The uni-ball is technically stronger, but you have to give up the adjustment of the rear suspension. Even though many builders use rear 1.25" rod ends on trailing arms without issues, opinions on this subject are mixed.

I opted for rod ends with adjustment for this truck, but in the future that could change. As for the front chassis mount on the lower arms, Iwent with something that is definitely considered to be a standard. It's a 1" uni-ball "wobble" stopper setup. These are machined by Blitzkrieg Motorsports and are designed to prevent the arm from twisting during movement. The worst thing you can have is a rear link arm that twists and rolls over through travel.

www.ingramcontent.com/pod-product-compliance
Lightning Source LLC
Chambersburg PA
CBHW082213290526
45794CB00009B/3528